ALLE ZEIT WACH
SJ
1842

ADVANCES IN BIOCHEMICAL ENGINEERING

Volume 5

Editors: T. K. Ghose, A. Fiechter,
N. Blakebrough

Managing Editor: A. Fiechter

With 31 Figures

Springer-Verlag
Berlin · Heidelberg · New York 1977

ISBN 3-540-08074-0 Springer-Verlag Berlin · Heidelberg · New York
ISBN 0-387-08074-0 Springer-Verlag New York · Heidelberg · Berlin

Library of Congress Catalog Card Number 72-152360.
Printed in Germany.

Typesetting, printing, and bookbinding: Brühlsche Universitätsdruckerei Gießen.

Contents

Editorial Guidelines

The aim of this series is to keep bioengineers and microbiologists informed of the fundaments and advances pertaining to the biochemical processes they need for the construction of bio-plants–be they for water purification, obtaining enzymes or antibiotics, for breeding yeasts, or those required for other special biochemical or biosynthetic operations. This series will likewise familiarize the biochemist with how the engineer thinks and proceeds in his work, as well as with the constructive aids at his disposal. Providing the various specialists with such extensive information is not an easy task: the backgrounds of the biochemist, the microbiologist, and the engineer are founded on entirely different bases; yet they must work side-by-side in the constantly changing field of biochemical engineering.
With this as foremost consideration, the Editors will make a special effort to present a selection of premises as well as new findings and ways of applying innovations that arise. The field of biochemical engineering is still developing and making advancements in highly industrialized nations; it is also becoming increasingly significant in those lands plagued by food shortages, which are still wrestling with problems of development today. Of primary interest for these countries are discoveries of methods for obtaining valuable natural substances and for disposing of wastes–where possible, recycling them into useful and even highly beneficial products. Advances in Biochemical Engineering can provide them with relevant contributions dealing with means of supplying food–proteins, in particular. Therefore, just as biochemistry and technology are brought together in this series, the reader will be offered contributions from industrial nations and from those countries that are presently in need of progress in the area of technology.
The Editors look forward to a strong influx of manuscripts and will do their utmost to insure the series' rapid publication. They will be published in English in order to afford the widest possbile outreach. Editors and Editorial Board are now prepared to accept manuscripts for consideration.

The Editors

Production of Cellulolytic Enzymes by Fungi

T.-M. ENARI and P. MARKKANEN
Technical Research Centre of Finland, Biotechnical Laboratory,
Box 192, SF-00121 Helsinki 12, Finland

Contents

Summary

Microorganisms able to utilize cellulose are found amongst bacteria, actinomycetes, and higher fungi. *Trichoderma viride* seems to be the best presently available organism for the production of extracellular cellulases. Most studies concerning the nature and mode of action of cellulases have also been carried out using this organism.

The enzymatic degradation of cellulose is a complex process requiring the participation of at least three types of cellulolytic activity: exo-β-1,4-glucanase, endo-β-1,4-glucanase, and β-glucosidase. In the hydrolysis of native cellulose exo- and endoglucanase act synergistically to produce cellobiose, which is then degraded to glucose by β-glucosidase. Some other enzymes may also be involved in the process, but definite evidence of this is lacking.

The synthesis of cellulase in *T. viride* is controlled by a repressor-inducer mechanism. The production of cellulases is thus greatly influenced by the carbon source in the medium. Glucose or other rapidly metabolized compounds cannot be used. Cellulose or some cellulosic material is probably the most suitable substrate for industrial cellulase production.

Cellulases are produced by surface culture methods, but the high price of the enzymes precludes their use in the biotechnical process industry. Research work aimed at developing industrial cellulase production by submerged fermentation has reached the pilot plant stage. The production is not economical at present, but continued research into improvement of microbial strains and process optimization may solve the problems in the near future.

1. Introduction

The food and energy shortages in the world have directed the interest of applied research workers toward the search for hitherto unused renewable resources. Cellulose is the major constituent of all plant material. It forms about one third of the woody tissues and is constantly replenished by photosynthesis. It is thus natural that a growing interest should be shown in the application of cellulolytic microbes and their enzymes to the utilization of cellulosic materials. The main use of extracellular cellulolytic enzymes would be in the hydrolysis of cellulosic materials in order to produce fermentable sugars for various biotechnical processes. One other important use would be in the treatment of fodder grain to increase its digestibility. Furthermore, cellulolytic organisms which do not excrete cellulases could be utilized for the production of single-cell protein by direct cultivation on cellulosic materials. A prerequisite for all technical applications of this kind is a thorough knowledge of the enzymes and their properties as well as the development of industrial processes for cellulase production.
Microorganisms producing enzymes hydrolyzing β-glucosidic linkages are widely distributed amongst various taxa. The ability to utilize cellulose is found amongst bacteria, actinomycetes, and higher fungi. The concept of cellulases can be limited to enzymes capable of degrading highly ordered cellulose into sugars small enough to pass through the microbial cell wall. It has been found that the degradation of cellulose is a complex process accomplished by the synergistic action of several enzymes. This review is confined to the enzyme complex hydrolyzing native cellulose to fermentable sugars. Since fungi are the organisms most likely to be used for industrial production of cellulases only fungal enzymes are discussed.

2. Nature of Cellulases

2.1 Cellulolytic Enzymes

The degradation of crystalline cellulose is a complex process, requiring the participation of many enzymes. It is now well established that there are at least three different types of cellulolytic activity: exo-β-1,4-glucanase (E. C. 3.2.1.–), endo-β-1,4-glucanase (E. C. 3.2.1.4), and β-glucosidase (E. C. 3.2.1.21). A strong synergistic effect has been observed between exo- and endoglucanases hydrolyzing crystalline cellulose (Avicel), but not when hydrolyzing acid-swollen cellulose [1]. β-Glucosidases hydrolyze cellobiose and short-chain cello-oligosaccharides to glucose, but have no effect on cellulose. Some β-glucosidases attack aryl-β-glucosides, but not cellobiose [2].
The first hypothesis concerning the nature of enzymatic hydrolysis of cellulose was put forward by Reese *et al.* [3]. They reported the existence of a nonhydrolytic enzyme, C_1, which initiated the hydrolysis of native cellulose by breaking hydrogen bonds between cellulose chains. This first step was a prerequisite for hydrolysis by hydrolytic enzymes,

C_x [4]. It was also believed that those microorganisms unable to grow on native cellulose did not synthesize C_1-enzyme. This model for the hydrolysis has subsequently been extensively questioned. In particular, the nature of the initial step in the hydrolysis of native cellulose is still obscure. At present the most generally accepted view is that C_1-enzyme is an exo-β-1,4-glucanase. In the case of *Trichoderma viride* and *T. koningii* purified C_1 has been shown to be a cellobiohydrolase [5–13]. Clearly we should now reconsider the theory of Reese and coworkers and redefine the mechanisms of cellulase action in the light of new understanding of the properties of cellulases. According to the present view, cooperative action of endo- and exoglucanases hydrolyzes crystalline cellulose to soluble cello-oligosaccharides, mainly cellobiose, which is released by exoglucanases [1, 11, 12, 14].

2.1.1 Exo- and Endoglucanases

Wood and McCrae [14, 15] separated the cellulase complex produced by *Trichoderma koningii* into eight pure components using gel filtration, ion exchange chromatography and isoelectric focusing. These components were a single exo-β-1,4-glucanase, C_1, five endo-β-1,4-glucanases, C_x, and two β-glucosidases. The complex thus contains many isoenzymes. The same authors reported that the exoglucanase was splitting off cellobiose from the non-reducing end of the cellulose chain [8, 11]. Thus, it may be systematically designated β-1,4-glucan cellobiohydrolase (E.C.3.2.1.–). The endoglucanases hydrolyze β-1,4-glucans in a random fashion and are systematically called β-1,4-glucan glucanohydrolases (E.C.3.2.1.4). The endoglucanases can be differentiated by the randomness of their attack on carboxymethyl cellulose (CMC) and by the rate of solubilization of phosphoric acid-swollen cellulose [8, 11].

Eriksson and coworkers studied the cellulase system of the rot fungus *Sporotrichum pulverulentum* (formerly called *Chrysosporium lignorum*) [1, 16]. Eriksson and Pettersson isolated five endo-β-1,4-glucanases and an exo-β-1,4-glucanase [16]. After isolation these proteins were found to be pure when tested using various methods. The same authors also quantitatively determined the ratio of activities between these five components to be 4 : 1 : 1 : 1 : 1.

Pettersson [12] fractionated the cellulase complex of *T. viride* into four components using gel chromatography, ion exchange chromatography, biospecific chromatography, and isoelectric focusing. Two of the components were endoglucanases, one was an exoglucanase, and one a cellobiase. The exoglucanase was shown to be a cellobiohydrolase, which was inhibited by cellobiose. Consequently β-glucosidase greatly accelerates the action of exoglucanase on microcrystalline cellulose by removing cellobiose.

It therefore seems clear that fungi produce at least five different endo-β-1,4-glucanases, the old C_x-components, varying in degree of randomness of hydrolytic action. So far only one exo-β-1,4-glucanase has been purified and fully characterized. It has, however, been shown clearly that all known organisms hydrolyzing native cellulose are able to produce at least one exo-β-glucanase. In the case of *T. viride, T. koningii* and *S. pulverulentum*, this enzyme is β-1,4-glucan cellobiohydrolase.

It has also been claimed that fungi produce a β-1,4-glucan glucosylhydrolase, but none of these enzymes has been isolated in a pure state. Preparations releasing glucose from cellulosic substrates have been isolated from culture media of *T. viride* [17] and *Aspergillus niger* [18]. However, these preparations were not pure enough to completely exclude the presence of cellobiase.

2.1.2 β-Glucosidases

The third activity involved in the breakdown of cellulose is β-glucosidase or cellobiase (E.C.3.2.1.21), which hydrolyzes mainly cellobiose, but also higher cellodextrins to glucose. These enzymes accelerate the hydrolysis of crystalline cellulose by removing cellobiose, which is an inhibitor of exo-β-glucanase. β-glucosidases are widespread in fungi. Bucht and Eriksson [2] isolated both β-glucosidase and aryl-β-glucosidase from *Stereum sanguinolentum. T. koningii* produces two β-glucosidases [11].

2.2 Properties of Cellulases

The molecular weights of the five endoglucanases isolated from *Sporotrichum pulverulentum* vary between 28 300 and 37 500 [16]. Small differences in the amino-acid composition have also been found. The isoelectric points vary between 4.20 and 5.32, making possible their separation by isoelectric focusing. With the exception of one component, all endoglucanases are glycoproteins. In Table 1 some properties of cellulases isolated from *T. viride* are summarized according to Pettersson [12]. The molecular weights of the exo- and endoglucanases of *T. viride, T. koningii, Fusarium solani* and *Penicillium funiculosum* lie in the region 40 000 ... 75 000, with the exception of the low-molecular-weight components from *T. koningii* and *T. viride.* These have a molecular weight of 12 500 ... 13 000 [12, 14].

Table 1. Some properties of cellulolytic enzymes isolated from *Trichoderma viride* [12]

Type of enzyme	Molecular weight	Isoelectric point	Carbohydrate content (per cent)	Activity toward different substrates			
				CMC	Microcrystalline cellulose	Reprecipitated cellulose	Cellotetraose
Exo-β-1,4-glucanase	42000	3.79	9	–	+	+	+
Endo-β-1,4-glucanase I	12500	4.60	21	+	–	+	+
Endo-β-1,4-glucanase II	50000	3.39	12	+	–	+	+
β-Glucosidase	47000	5.74	0	–	–	–	+

Thermostability is one of the most important technical properties of cellulases, since the hydrolysis of cellulose proceeds faster at higher temperatures. Endoglucanases are more stable than exoglucanases. Endoglucanases are quite stable for up to 4 hrs at 60° C and pH 5.0. β-Glucosidase and exoglucanase of *T. koningii* resemble one another in their heat stability at 60° C: they loose about 80% of their original activity at 60° C and pH 5.0 in 4 hrs [14]. In the presence of cotton the cellulases of *T. koningii* and *F. solani* are remarkably stable, showing no loss of activity when incubated for 4 weeks at 37° C and pH 5.0 [14].

2.3 Mode of Action of Cellulases

Wood and McCrae purified the exoglucanase (C_1-component) of *Trichoderma koningii* using ion exchange chromatography on a DEAE-Sephadex column and pH gradient elution [8, 14, 19]. The low-molecular-weight endoglucanase, the removal of which does not affect the kinetics of solubilization of cotton fiber [20], was first separated from the culture filtrate by gel chromatography on a Sephadex G-75 column. The remaining fraction containing endoglucanases (C_x) and β-glucosidases, was separated according to the scheme in Fig. 1.

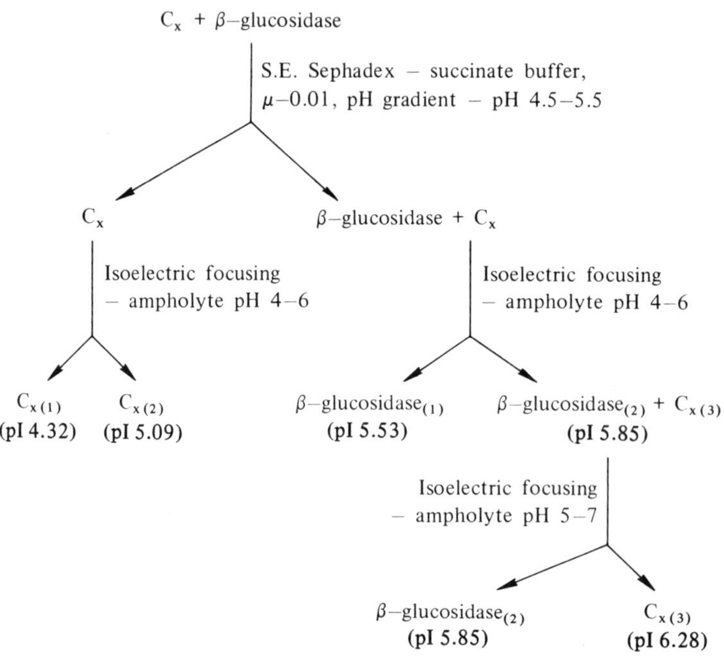

Fig. 1. Wood's and McCrae's scheme for fractionation of cellulolytic enzymes from *Trichoderma koningii* [11]

The synergistic properties of the separated enzymes were studied by the same authors [11, 14]. They could reconstitute the cellulase complex from the fraction because the recoveries of the enzymes and protein were very high, over 90% [11]. Table 2 shows some of the reconstitution results.

Table 2. Relative cellulase activities of the components of *Trichoderma koningii* cellulase alone and in combination [11]

Enzyme	Relative cellulase activity (%)
C_1	< 1
$C_{x(1)}$	< 1
$C_{x(2)}$	< 1
β-Glucosidase$_{(1)}$	0
β-Glucosidase$_{(2)}$	0
$C_1 + C_{x(1)} + C_{x(2)}$	24
C_1 + β-glucosidase$_{(1+2)}$	5
$C_1 + C_{x(1+2)}$ + β-glucosidase$_{(1+2)}$	103
20–80% sat. $(NH_4)_2SO_4$ fraction	100

All of the original cellulase activity was reconstituted when all the components, C_1 + $C_{x(1+2)}$ + β-glucosidase$_{(1+2)}$, were recombined in their original proportions. The most potent synergistic effect was found between exoglucanase (C_1) and the endoglucanase component ($C_{x(2)}$) when cotton was used as substrate. $C_{x(1)}$ and C_1 act synergistically on cellulose, but the low-molecular-weight C_x-component showed no synergistic effect at all. Glucose and cellobiose were the main products when exo- and endoglucanases were combined. However, the proportion of glucose was low: 8% when the combination $C_1 + C_{x(1)}$ was used, and 14% for $C_1 + C_{x(2)}$.

Wood also observed other differences in the hydrolytic capacities of *T. koningii* cellulases. The endoglucanase, $C_{x(1)}$-component, hydrolyzed 29% of phosphoric-acid-swollen cellulose in 4 h, while the endoglucanase, $C_{x(2)}$, hydrolyzed 83%, and exoglucanase, C_1, 32% [11]. The endoglucanase, $C_{x(2)}$-component, is more random in its action than $C_{x(1)}$. Obviously, therefore, the combination of exoglucanase with the endoglucanase, $C_{x(2)}$-component, hydrolyzes cotton cellulose more efficiently; endoglucanase, $C_{x(2)}$, opens more end groups for the action of exoglucanase. This finding supports the theory that endoglucanases initiate the attack on native cellulose.

Exo-β-glucanase from *S. pulverulentum* showed no viscosity-decreasing activity toward CMC [13]. The main product, cellobiose, is released in the α-configuration [1]. Eriksson and Pettersson [16] found that the weight ratio of exoglucanase protein to endoglucanase protein was 1 : 1. They also found a strong synergistic action between exo- and endoglucanases when hydrolyzing crystalline cellulose, but not when hydrolyzing phosphoric-acid-swollen cellulose [1]. Endoglucanase pretreatment also increased the production of cellobiose from cotton cellulose by exoglucanase. This also supports the theory that the endoglucanases open chain ends for exoglucanases.

This evidence strongly supports the mechanism for enzymatic degradation of cellulose as described by Pettersson [12] (Table 3).

In the first reaction free ends for exoglucanase are released at the sites of non-crystalline regions of the cellulose. This kind of mechanism was first suggested by Eriksson [21, 22]. Different endoglucanases have different substrate specificities and therefore can attack a variety of substrates.

Table 3. A mechanism for enzymatic cellulose degradation [12]

1. Native	cellulose	Endoglucanase →	Cellulose[a]
2.	Cellulose[a]	Exoglucanase →	Cellobiose
3.	Cellobiose	β-glucosidase →	2 Glucose

[a] Formed from native cellulose by the action of the endoglucanase on non-crystalline regions of the cellulose fiber. Free chain ends are created.

The initiation mechanism of the degradation of cellulose has not yet been completely clarified. It is still possible that hitherto unknown enzymes are involved in the degradation. The existence of one such enzyme was demonstrated by Eriksson and coworkers [23]. The quantitative purification of both exo- and endoglucanases from culture filtrate of *S. pulverulentum* made it possible for them to reconstitute the culture solution using purified enzymes.

The reconstituted solution contained the same quantities of endo- and exoglucanases as the original culture solution. The concentrated unfractionated culture solution degraded 52.1% of de-waxed cotton, whereas the reconstituted solution degraded only 20% [13, 23]. They believed that an additional enzyme important for the degradation of crystalline cellulose was present in the culture solution, but not in the reconstituted solution. When the culture solution was incubated with nitrogen instead of air, the degree of cellulose degradation decreased from the original 52.1% to 21.5% (Table 4). This indicates that there is an additional oxidizing enzyme involved in the degradation of cellulose. It was also shown that the same enzyme is present in the culture solutions of other cellulolytic fungi (Table 5). Eriksson has suggested that the probable mode of action of this oxidizing enzyme comprises insertion of uronic-acid moieties into the cellulose, thus breaking the hydrogen bonds between chains. However, the enzyme has not yet been characterized and purified in sufficient amounts for a final verification of this.

Table 4. Degradation of cotton cellulose by enzymes from *Sporotrichum pulverulentum* [23]

Tube No.	Enzyme preparation	Cellulose degradation, weight loss %
1	Concentrated culture solution	52.1 (oxygen atmosphere)
2	Concentrated culture solution	21.5 (nitrogen atmosphere)
3	Mixture of endo- and exoglucanases	20.0
4	Endo-β-1,4-glucanases	0.0
5	Exo-β-1,4-glucanase	0.0

Most of the studies concerning the degradation of cellulose have been made using pure cellulose as substrate. In natural materials cellulose is usually present as a complex. *S. pulverulentum* produces the enzyme cellobiose: quinone oxidoreductase which participates in the degradation of cellulose in wood [13, 24, 25]. This enzyme needs a quinone as a cosubstrate (quinones are released from lignin) and therefore cannot function in the degradation of pure cellulose. Thus, degradation studies with pure cellulose as substrate may be misleading.

Table 5. Degradation of cotton cellulose by cell-free, concentrated culture solutions of four different cellulose-degradating fungi in presence and absence of oxygen [23]

Organism	Cellulose degradation (weight loss %)	
	O_2-atmosphere	N_2-atmosphere
Sporotrichum pulverulentum[a]	52.1	21.5
Polyporus adustus[b]	42.6	18.0
Myrothecium verrucaria[b]	33.6	17.0
Trichoderma viride[c]	20.0	10.0

[a] Culture solution concentrated 50 times.
[b] Culture solution concentrated 30 times.
[c] Culture solution concentrated 20 times.

It is apparent, therefore, that the degradation of cellulose is a complicated process. It has been clearly demonstrated that there is a synergistic effect between exo- and endoglucanases. Furthermore, β-glucosidase is needed for the removal of cellobiose, which otherwise inhibits the action of exoglucanase. The oxidizing enzyme observed by Eriksson [13] may be involved in the degradation of crystalline cellulose, but its function has yet to be demonstrated. It is also likely that new kinds of cellulolytic enzymes will be discovered in the near future, as there is a great amount of research activity in this field. It is tempting to replace the old C_1–C_x concept by more precise names, such as exo-β-glucanase and endo-β-glucanase. However, confusion may arise through replacement of the term C_1 by exoglucanase, since there are also exoglucanases which do not attack insoluble cellulose [26]. The mechanisms of the action of the cellulase complex may also be very different in different organisms [27].

2.4 Activity Determinations

Determination of the activities of cellulolytic enzymes is complicated by two factors: 1. In most cases determinations are not made on purified enzymes, but rather on solutions containing a mixture of different cellulolytic enzymes. Because of the synergistic action of these enzymes, the activity measured is greatly influenced by the proportions of different enzymes, which may vary; 2. The substrates used are natural macromolecules, which makes standardization difficult. The ideal substrate would be of low molecular weight and specific. Unfortunately, only in the case of β-glucosidase such a substrate is available.

In developing methods for activity determinations, two different approaches can and have been adopted. In the technical approach the starting point is the use of the cellulases. The main technical use of cellulases is to produce glucose from various cellulosic materials. Hence, this approach leads to a method in which the substrate is a suitable cellulosic material (e.g. filter paper) and the end-product formed, glucose, is measured. Methods of this type give a value for the overall cellulolytic activity, but give no indication of which enzyme is rate-limiting. They are useful for determining the capacity of a certain enzyme complex to hydrolyze cellulose, but they are not methods for the determination of the activity of individual enzymes.

In the biochemical approach the aim is to determine the activity of individual enzymes. These methods are necessary for research into the biochemical mechanism of enzymatic cellulose degradation. They are also very useful in screening cellulase-producing organisms and in developing enzyme production processes. Thus, measures can be taken to improve the limiting activity. The difficulty in developing methods for individual activities is that it is necessary to know which enzymes are involved in cellulose degradation. Thus, a considerable amount of biochemical research is necessary in the development of such methods. Another difficulty is the lack of specific substrates or inhibitors which would permit measurement of one activity in the presence of other synergistic activities.

2.4.1 Overall Cellulolytic Activity

In determinations of the overall cellulolytic activity, the substrate must resemble the one which will be used in a technical hydrolysis process, i.e. it must be an insoluble cellulosic material which is not too easily hydrolyzed. It must, nevertheless, be a material which can be standardized. Another important factor is the reaction time. Since the substrate is an insoluble fibrous material, time is required for the enzyme to diffuse into the fiber and for the hydrolysis products to diffuse out of the fiber. Another difficulty is caused by the varying accessibility of glucosidic bonds in different regions of the fiber. In order to give a meaningful result, the assay requires a reaction time long enough for hydrolysis of an appreciable fraction of the less accessible bonds. Thus, for overall cellulolytic activity the generally accepted rule of measuring the initial reaction rate cannot be followed.

Various substrates have been proposed. Cotton fiber is one of the most resistant. Avicel, a microcrystalline cellulose, is also difficult to hydrolyze. Sulphite pulps, such as Solka Floc and filter paper, have also been used. Filter paper has proved to be a satisfactory substrate for the measurement of overall cellulolytic activity. The method of Mandels and Weber [28] has gained general acceptance for this purpose. In this method the reducing sugars formed under standard conditions are estimated. The reaction time used is relatively short, one hour, and hence the measurement is based on limited action of the enzymes on the most susceptible regions of the substrate. Increasing the enzyme or substrate concentration leads to increased sugar production, as does a prolonged reaction time. The increase in glucose formation is not linear because the most reactive substrate is converted at the beginning of the reaction period [29]. The activity values are erratic at high glucose concentrations and tend to become less meaningful with highly active preparations. Linearity can be increased by diluting the enzyme, increasing substrate concentration or decreasing the reaction time [30]. It has been suggested that the most reliable quantitative activity determinations ought to involve enzyme units based on the same degree of hydrolysis of the filter paper, e.g. a dilution giving 2 mg of glucose [29] under the test conditions.

The most difficult step in the technical hydrolysis of cellulosic materials is the solubilization of fibrous substrates. It is therefore natural that special emphasis be placed on determination of the solubilizing activity, which has often been referred to as C_1-activity.

It has been shown that solubilization is caused by the synergistic activity of exo-β-glucanase and endo-β-glucanase [30]. Methods for determination of the solubilizing activity make use of cotton fibers, microcrystalline cellulose (Avicel), or hydrocellulose as substrate, with measurement of the production of reducing sugars [28], the loss of weight [31] or the decrease in optical density of a cellulose suspension [32].
When the formation of reducing sugars is measured, the acitivity determined is the sum of different cellulolytic activities, and the result depends on the relative proportions of the different enzymes. The formation of cellobiose or glucose as the end-product depends on the β-glucosidase activity, which can thus greatly influence the result [33]. Determination of the nonsolubilized substrate after enzymic digestion gives a reliable result, but the method is laborious and unsuitable for long series of determinations. Measurements based on the reduction in optical density of cellulose suspensions are useful for the screening of cellulase-producing organisms or mutants. In a plate assay, cellulase-producing organisms can be detected by formation of a clear zone when growing on a medium containing phosphoric-acid-swollen cellulose [34, 35].
The release of dye from a dyed insoluble substrate is a convenient way of measuring the solubilizing activity. Dyed filter paper [36], dyed Solka Floc [37], and dyed Avicel [37] have been used as substrates. There is also a commercial substrate available, Cellulose azure (Calbiochem, Switzerland). The best substrate for determining the solubilizing activity is dyed Avicel SF [33, 37]. Dyed Solka Floc and the commercial Cellulose azure are too easily solubilized. The method using dyed Avicel is convenient and rapid. It is thus a good tool for research into the production of cellulases and screening of cellulolytic microorganisms.
A number of methods making use of less well-defined activities have also been proposed. Such methods include swelling of cotton or paper, maceration of paper, decrease in breaking strength of yarn, thread, or fabrics, and microfragmentation of cellulose micelles [28].

2.4.2 Endo-β-Glucanase

Endo-β-1,4-glucanases randomly attack β-1,4-linkages in CMC or swollen cellulose. The best substrate for the measurement of endo-β-glucanase activity is a soluble cellulose derivative such as CMC. This substrate has been employed by many workers, who measured either the decrease in viscosity [38] or the production of reducing sugars [28]. Measurement of the decrease in viscosity is a very sensitive technique, since even a few breaks in a chain cause a marked decrease in the average chain length. Measurement of the reducing sugars is less sensitive and is also influenced by the presence of other cellulolytic enzymes, especially β-glucosidase. CMC is not, as such, attacked by cellulolytic enzymes other than endo-β-glucanase [12] (Table 1), but the cellobiose formed is, of course, hydrolyzed by β-glucosidase. Ionic-substituted celluloses, for example CMC, are not the ideal substrates for viscometric assays. Difficulties are caused by the fact that the viscosity of ionic substrates is dependent on pH, ionic strength, and polyvalent cations. For this reason, non-ionic-substituted celluloses, such as hydroxyethyl cellulose (HEC), are preferred for the determination of low endo-β-glucanase activities [39].

2.4.3 Exo-β-Glucanase

Exo-β-glucanase is the first enzyme involved in the breakdown of insoluble cellulose preparations. Consequently these are suitable substrates for the determination of exo-β-glucanase. Usually cotton is considered the best substrate [28], but micro-crystalline cellulose is also suitable. Microcrystalline cellulose is hydrolyzed only by exo-β-glucanase (Table 1). Since the enzyme produces cellobiose as the reaction product, the measurement of the reducing sugars formed is influenced by the presence of β-glucosidase. On the other hand, endo-β-glucanase action opens up new chain ends, producing more substrate for the exo-β-glucanase. If the reducing sugars produced from microcrystalline cellulose or cotton are estimated, the measurement is a true determination of exo-β-glucanase only when other cellulolytic enzymes are absent.

2.4.4 β-Glucosidase

Fewest problems arise in the activity determinations for the enzyme hydrolyzing cellobiose. This enzyme hydrolyzes both cellobiose and β-1,4-oligosaccharides (Table 1) to glucose. It can be determined using cellobiose as a substrate [40]. β-Glucosidase can also utilize a pseudosubstrate, p-nitrophenyl-β-glucoside [20], the use of which provides a rapid and convenient determination method.

3. Microbial Sources of Cellulases

3.1 Producers of Extracellular Cellulases

The ability to produce cellulolytic enzymes is widespread amongst microbes. The selection of a cellulase-producing organism depends on the purpose for which the enzymes are intended. For industrial production the most efficient cellulase producer must be selected. For research or other laboratory purposes the amount of cellulase produced may be less significant.

The availability of highly active cellulase preparations is a prerequisite for the industrial production of glucose from cellulose. *Trichoderma viride* seems to be the best presently available source of extracellular cellulases. Many other cellulolytic organisms have been studied and it may be possible to obtain good cellulase yields using them. Many bacteria and fungi that are able to grow on native or pretreated cellulose do not, however, secrete cellulases into the medium. Some microbes secrete only endoglucanases (C_x) or β-glucosidase, and they are therefore unable to hydrolyze native crystalline cellulose. Only true cellulolytic organisms possessing exoglucanase activity (C_1) can hydrolyze native cellulose. Such organisms are found amongst the higher fungi.

Only a few fungi have been reported to produce cellulases degrading native cellulose. Such fungi are: *T. viride, T. lignorum* and *T. koningii* [7–9, 22, 28, 40–43], *Sporotrichum pulverulentum* [1, 13, 16], *Penicillium funiculosum* [44] and *P. iriensis* [45], *Polyporus adustus* [13], *Myrothecium verrucaria* [46], *Fusarium solani* [7], and *Chaetomium thermophile* var. *dissitum* [27].

Many more fungi and bacteria produce cellulases which degrade pretreated cellulose or carboxymethyl cellulose (CMC), but not crystalline cellulose [47, 48]. Cellulose-degrading forms are found among the gliding bacteria, among Gram-negative and Gram-positive true bacteria, and among actinomycetes [27, 49]. Cellulolytic ability is also found among obligate aerobes (gliding bacteria, *Pseudomonas*), facultative anaerobes (*Bacillus, Cellulomonas*), and obligate anaerobes (*Clostridium*).

In recent years thermophilic organisms have also been studied. The ascomycete, *Chaetomium thermophile* var. *dissitum*, is a typical thermophilic fungus able to produce a cellulolytic system decomposing native cellulose [27]. *Chaetomium thermophile, Sporotrichum thermophilium,* and *Thermoascus aurantiacus* grow on and decompose cellulose very rapidly, but the cellulase activities of the culture filtrates are low [29]. *Thermomonospora curvata* has been observed to produce both endo- and exoglucanases when grown on cellulose [50]. The enzyme system secreted by this organism caused less than 1% hydrolysis of cotton fiber, indicating that it is unable to decompose native cellulose. Interest in thermophilic organisms has been stimulated by the search for thermostable cellulases. However, cellulases from thermophiles are not necessarily more heat-stable than cellulases from mesophiles [29]. Mandels [29] compared the cellulase systems produced by *Thermoactinomyces* and *Trichoderma viride.* In short assays on susceptible substrates both cellulases showed higher activity at 65° C than at 50° C. In a 24-hrs assay on cotton the *T. viride* cellulase was inactivated at 60° C, but *Thermoactinomyces* cellulase was found to be deficient in exoglucanase activity, so further comparison could not be made.

Although much work has been done with other organisms, *T. viride* still seems to be the most convenient source of extracellular cellulases. One possibility of attaining high overall cellulolytic activities lies in mixing the enzymes produced by different organisms. However, the mechanisms of cellulose decomposition may be different in different microbes. Not all endoglucanases act synergistically with all exoglucanases [27]. For synergism, two enzymes must work together in the form of a loose complex, which cannot be formed between all exo- and endoglucanases.

3.2 Other Cellulolytic Organisms

Many organisms degrade cellulose by direct contact with the substrate, and the occurrence of cellulase activity in culture supernatants is often due to autolysis. It seems in many cases that cellulases are most efficient when cell-bound. At the cell surface the enzymes occur in high concentration and can achieve close contact with the substrate. The enzyme-substrate complexes formed along the crystalline cellulose fiber may allow the change in conformation of cellulose needed to make it susceptible to hydrolysis [51]. In the degradation of cellulosic materials syntrophism, in which two microorganisms can grow together on a given substrate whereas each alone cannot, is often observed [52]. Symbiotic growth of cellulolytic organisms, mainly *Cellulomonas*, and cellobiase-producing organisms is often used to increase the biomass yield [53]. Pretreatment of cellulose enhances the growth of *Cellulomonas* because the number of organisms adhering to the fibers during fermentation increases. The bacteria are arranged in a regular manner along the surface of the cellulose fiber [53]. A similar phenomenon was observed by Berg *et al.* in the degradation of cellulose fibers by *Sporocytophaga myxococcoides* [54]. The secretion of enzymes into the medium is not the only way to utilize cellulolytic enzymes. SCP-production by direct cultivation on cellulosic materials may in the near future become as important as the production of extracellular cellulase.

4. Production of Cellulases

The economical production of cellulases depends on the selection and improvement of suitable strains and on the development of fermentation media and methods. Most of the research concerning production methods has been done using *Trichoderma viride* strains. One great difficulty is the low productivity of the strains and the long time needed for cultivation. Mutagenesis of *T. viride* has produced only a three- to fourfold improvement in cellulase yields. This low hyperproduction of cellulases compared to the hyperproduction of some other fungal enzymes (amylases, proteases) may be a result of the induction mechanism of cellulases. Catabolite repression is another regulatory mechanism by which the concentration of cellulases is regulated. Improvements in fermentation methods and the search for regulatory mutants are the most promising means for maximizing cellulase production.

Some commercial *T. viride* enzyme preparations are available, but for a high price [55, 56] (cf. p. 20). There is a great deal of research activity throughout the world aimed at the development of economical methods of cellulase production. Much of this work is being carried out in the U. S. Army Natick Laboratories, Natick, Mass., by Reese, Mandels, and coworkers. High-producing strains of *T. viride* have also been isolated and improved there.

4.1 Cultivation Conditions

Trichoderma species produce various carbohydrases besides cellulases. Among them are xylanase and mannanase. Waste cellulosic materials contain various types of carbohydrates. When enzymes are produced for use in the hydrolysis of such materials it may be desirable to use the same waste material as a carbon source to induce the proper mixture of enzymes.

4.1.1 Media

The basic media for growth and cellulase production by *T. viride* have been described by Mandels and coworkers [28, 41]. The media contain peptone (0.05–0.1%) and urea as nitrogen sources, different cellulose preparations (0.5–1.5%) as carbon source, necessary minerals, and 0.2% Tween 80. It is well known that surface active agents stimulate the production of extracellular enzymes [57].

As the production of cellulases is inducible [29, 58–61] the production of cellulases is greatly influenced by the nature of the carbon source. Because catabolite repression by glucose or other readily metabolizable compounds also controls the production, care must be taken in the choice of carbon source. Different pretreated celluloses are the cheapest carbon sources.

Peptone, used as nitrogen source by Mandels and coworkers [58, 59, 62], is too expensive for industrial use. Organic nitrogen is, however, needed for maximum enzyme production. Peptone can be replaced by some industrial waste materials and also by *Trichoderma* cells from previous cultures. The precise choice depends on local conditions. Inorganic nitrogen compounds may also be used. These are more suitable for growth than for enzyme production. The nitrogen source influences the pH, and because the pH is a very important factor affecting enzyme production the choice and concentration of nitrogen source is very important.

4.1.2 Enzyme Production and Growth

One difficulty in monitoring the cultivation of mycelial organisms is the measurement of growth. In the production of cellulase undissolved cellulose continuously decreases while biomass increases. Therefore, in most cases only enzymatic activities have been recorded. Acid production is directly related to the rate of carbohydrate consumption. As soon as all carbohydrate is consumed, the pH rises by secretion of ammoniacal compounds or by consumption of acids formed during previous growth. A typical enzyme production curve is illustrated in Fig. 2.

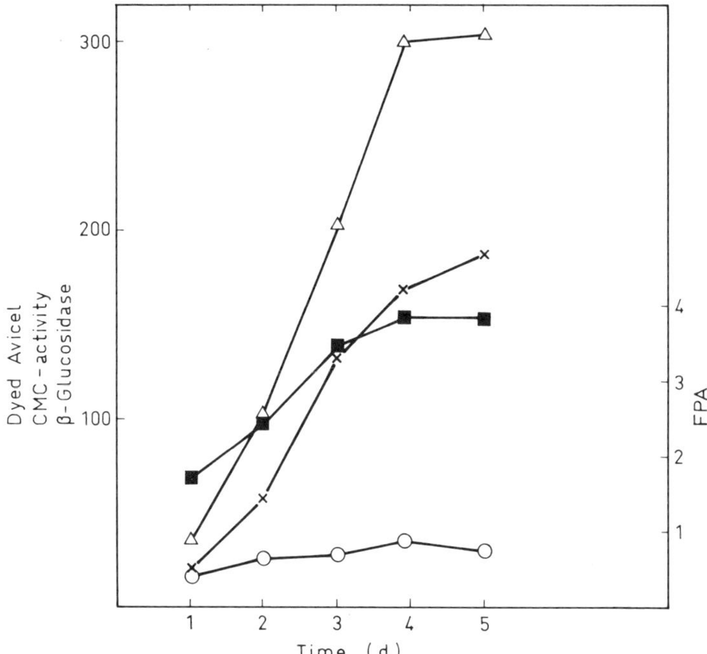

Fig. 2. Cellulolytic enzyme activities from a typical batch fermentation of *Trichoderma viride*. △ Activity against dyed Avicel, × Activity against CMC, ○ β-Glucosidase, ■ FPA

During the acid-production stage, when metabolic activity is high, cellulases are induced. Acid production also seems to have a regulatory function [56]. If glucose is added to cultures growing on cellulose, the pH drops to 2.5, with a significant loss of cellulase activity. If an inducer is still present after the glucose is consumed, the enzymes reappear [63]. If the pH is held at about 5.0 after glucose addition, the activity of enzymes remains stable. Thus, the observed apparent glucose effect is in fact related to pH [56]. Brown *et al.* [64] showed the feasibility of producing cellulase from *T. viride* with commercial glucose as the only carbon source. This could be due to the fact that glucose produced by acid hydrolysis contains sophorose, an inducer of cellulase. The production of cellulases was repressed by glucose, and glucose had to be absent from the medium before the enzymes could be produced. A low pH was, however, necessary to slow down metabolic activity. After exhaustion of glucose, the cells require a period of maturation before cellulase production begins. Brown *et al.* also succeeded in producing cellulase in continuous culture. Mandels [29] reported that she was unable to obtain even the modest levels of cellulase obtained in shake flasks when using glucose as the substrate for growth. Brown *et al.* have stated that complex interactions exist between the medium composition, pH, inoculum size and state, and aeration capacity.

4.1.3 Pilot Plant Investigations

In pilot plant experiments Nystrom and Kornuta [62] have stated that the fermentation profile is complicated by the complex media and the type of inoculum used. The size and stage of growth of the inoculum controls the initial lag phase of growth. Inoculum sizes of up to 10% are preferable and a high pH promotes rapid growth [62, 65]. The inocula must themselves contain cellulolytic activity in order to facilitate initial growth. If suitable conditions are used, the growth rate becomes controlled by the release of sugars. Nystrom and Kornuta followed shifts in metabolism by measuring the respiratory quotient.

Foam production is problematic in many enzyme fermentations. The production of cellulases needs, however, only a low dissolved oxygen concentration. A volumetric adsorption coefficient ($k_L a$) of 50 mmol O_2 $1^{-1} h^{-1} atm^{-1}$, which maintains dissolved oxygen levels of 15% saturation, is sufficient for cellulase production. The relatively low aeration rates that are needed make foam control easier [62]. Using conditions given by Nystrom and Allen [65] the time for maximum enzyme production can be reduced to 3 to 4 days.

Taking into account the maximum time of 3 to 4 days for maximal enzyme production, Nystrom and Allen [65] proposed a scheme for a production line for cellulases using *T. viride.*

The operation of this line is semicontinuous. There are four lines, each containing four vessels. The first vessel is operated until the fermentation is 25 hrs old, when 10% of the culture is transferred as inoculum to the next. The second vessel is cultivated for 25 hrs, and again 10% of its culture is used as inoculum for the third vessel; and so the procedure continues in each line. The first fermentation is complete after 90 hrs. The vessel is then harvested, sterilized, and refilled. After 100 hrs the fourth vessel is ready for transfer of 10% of its 25-hrs old culture to the first vessel. In the case of contamination or other shortcomings, an inoculum can be taken from a vessel in another line, both lines continuing uninterrupted. After Kornuta this kind of system has been named the "Kornuta merry-go-round".

4.2 Induction and Repression

Cellulase is an inducible enzyme complex in *Trichoderma viride* [58–60, 63], but has been shown to be constitutive in *Pseudomonas fluorescens* [47]. Cellulase is produced when *T. viride* is grown on cellulose, lactose, glucose, and cellobiose [60]. Glucose, however, does not appear to be an inducer, since a high initial concentration is required and the synthesis of cellulase begins only after glucose is totally exhausted from the medium. The slight inducing effect of industrial glucose is presumably caused by sophorose formed during acid hydrolysis of starch [58, 59, 64]. Sophorose (2-0-β-D-glucopyranosyl-D-glucose) was found by Mandels *et al.* to be a potent inducer [66]. However, it has not been proved that sophorose would be involved in the synthesis of cellulase in natural conditions. Cellulose is converted mainly to cellobiose and glucose during the hydrolysis with enzymes. Cellobiose, in turn, markedly stimulates the production of cellulase if the fungal growth is restricted by culturing under suboptimal conditions [63]. It has been assumed that cellobiose is the natural inducer in cellulase synthesis.

The activities of cellulase and aryl-β-glucosidase were markedly increased by shaking washed cells of *T. viride* with sophorose [58, 59]. A similar effect was observed with gentiobiose, but the stimulation was very weak. Insoluble cellodextrin and cello-oligosaccharides only had a slight effect. The optimum concentration of sophorose was found to be 10^{-3} M, while concentrations higher than 10^{-1} M inhibited the formation of cellulase. This is clearly due to hydrolysis of sophorose by β-glucosidase. Thus glucose, which is a real repressor, is formed. When growing cultures with glycerol as the main carbon source no enzyme was synthesized without sophorose. When sophorose was added the synthesis began after the glycerol was exhausted. The synthesis of cellulase induced by sophorose is totally repressed by addition of 10^{-2} M glucose.

Crystalline cellulose can be used as the sole carbon source for cultures [60, 63, 67]. In washed cells neither cellulose nor cello-oligosaccharides, including cellobiose and cellodextrin (the degradation products of cellulose), induce the formation of cellulase [58]. Sophorose is not a degradation product of cellulolysis and, therefore, if it is a natural inducer, it must be synthesized during the induction phase.

It has also been shown that cellulase from *T. viride* possesses transglycosylation activity [68] which is needed in the synthesis of sugars like sophorose.

Nisizawa *et al.* [59] studied the enzyme induction caused by sophorose. Sophorose enhanced the formation of xylanase as well as that of cellulases and β-glucosidase. In a control experiment without sophorose, only β-glucosidase and xylanase were synthesized, but sophorose addition considerably increased the formation of these enzymes. Cellulase was synthesized only in the presence of sophorose. Furthermore, L-leucine-^{14}C was incorporated into cellulase protein in the presence of sophorose, but not in its absence, showing that sophorose also causes *de novo* cellulase synthesis. It was shown that most of the increase in glucose-producing enzyme activity by sophorose is due to the enhancement of exoglucanase activity and that the cell-bound β-glucosidase is an ordinary constitutive aryl-β-glucosidase, which is not important in cellulose breakdown [13]. Nisizawa and coworkers concluded that sophorose triggers the induction of *de novo* synthesis.

The formation of catabolic enzymes is usually repressed by glucose and other rapidly metabolizable compounds. The real chemical nature of the repressor or the mechanism of the catabolite repression is not exactly known. Sophorose-induced formation of cellulase in *T. viride* was strongly repressed by 10^{-2} M glucose [61]. The concentration of sophorose had an effect on the induction, but high concentrations of glucose (10^{-1} M) repressed even the maximum induction. Fructose, maltose, gluconate, some acids of the citric acid cycle, and ATP also caused catabolite repression [58].

In order to study the mechanism of catabolite repression, the effects of puromycin and actinomycin D were compared with the glucose effect [61]. The inhibitory effect of puromycin and glucose on washed cells of *T. viride* was complete within 30 to 45 min after the addition, whereas 60 to 90 min was needed for the effect of actinomycin D. Cellulase concentration reached a higher level when actinomycin D was added than after puromycin or glucose addition. Since puromycin inhibits the protein synthesis at the translational level it seems that glucose exerts its effect on the cellulase synthesis at the same level.

In conclusion, the formation of cellulase in *T. viride* is controlled by a repressor-inducer mechanism. The most efficient inducer is sophorose. The inductive formation of cellulase by sophorose is strongly repressed by glucose and other readily metabolizable compounds. The synthesis is clearly inhibited at the translational level. It has been suggested that cellobiose is the *in vivo* inducer of the cellulase synthesis, but this question still remains open. Although cellobiose at low concentrations stimulates the production of cellulase [69], its precise role in the induction has not been clarified. In growing fungal cultures cellulases are also induced directly or indirectly by the products of their action [56, 63, 69].

4.3 Genetic Improvement

In order to improve the enzyme yields of fungal fermentations the regulatory mechanism of enzyme synthesis must be known. In many cases the hyperproduction of an enzyme in a mutant can be up to 100 times that of wild strain. The synthesis of cellulases is controlled by catabolite repression and induction. De-repression occurs when the concentration of repressor is low or when it is totally exhausted from the medium. After that, the presence of an inducer is necessary. The reason that only three- to fourfold enhancement of production has been obtained by mutagenizing *Trichoderma viride* [70, 71] may lie in the inducibility of cellulases. Another reason for the small enhancement may be related to the quantity of enzyme protein needed for the hydrolysis of cellulose. The hyperproducers of cellulase secrete about 2 mg/ml of protein [61], which is already sufficient to eliminate any possibility of its being increased several times. Unfortunately, this protein is only partly cellulase. It seems, therefore, that in the search for mutants new aspects must be considered. Sternberg [70] gives some examples of the kind of mutants which should be looked for:

1. Control mutants, such as constitutive cellulase producers.
2. Mutants with minimized production of proteins other than cellulases, allowing more protein precursors for cellulase production.
3. Once the nature of the complex of cellulase enzymes is understood, it may be possible to obtain mutants which hyperproduce particular enzymes of the complex, and a more active complex would be obtained by mixing different culture filtrates.

One method to enhance enzyme synthesis is to produce heterokaryons [72], or to use transformation in which short DNA fragments are transferred. Thus, the number of genes for cellulase may increase.

The most used agents in the mutation of microorganisms are: alkylating agents, such as N-methyl-N-nitro-N-nitrosoguanidine (NG), diethyl sulfate (DES), and nitrous acid (HNO_2), as well as gamma- or UV-irradiation [35]. Spore suspensions are exposed to mutagenic irradiation or chemicals at doses killing > 90% of the spores. After mutagenic treatment the cells are diluted and grown on a complete medium forming separate clones of mutant strains. The mutated, cloned, and purified isolates are then cultivated on selective media. After this, prominent strains are inoculated into shake flasks containing a medium suitable for cellulase production. The lack of an effective plate assay renders rapid detection of high cellulase mutants impossible [71]. Some methods are, however, available for preliminary tests [35]. Nevertheless, the only effective way of screening involves cultivation of each isolate in shake flask and assay for cellulases. In order to find constitutive mutants cultivation on a glucose-containing medium is necessary.

The initial improvement in cellulase production induced by mutation is usually good, but further improvement becomes progressively more difficult [35]. In Table 6 some results are summarized for the improvement of cellulase production by *T. viride* strains according to Palva and Nevalainen [35]. The yield of different activities is usually only 3 to 4 times higher than that of wild strains [35, 56, 71, 72].

In conclusion, it can be stated that in order to improve the results of genetic manipulation of cellulase producers it is necessary to develop more specific selective techniques and enzyme assay procedures. This would simplify the preliminary screening of large

Table 6. Origin and enzyme activity of mutant strains isolated from strain VTT 304 [35]

Strain	Origin and treatment	Activity against soluble cellulose		Activity against insoluble cellulose	
		units/ml	%	units/ml	%
QM 9414	QM 9123 γ-rad.	240	100	60	100
VTT 304	QM 9414 spont.	280	117	68	113
A	VTT 304 NG	310	129	80	133
B	VTT 304 NG	310	129	77	129
C	VTT 304 DES	400	166	92	153
D	VTT 304 NG	350	146	92	153
E	VTT 304 NG	340	141	71	131

numbers of isolates. It may also be possible to transfer genetic information using transformation or other genetic methods. This would increase the gene dosage and thus cause enhancement in enzyme production. Constitutive mutants able to produce cellulase on glucose are also an interesting possibility. Unfortunately they are difficult to obtain since the appearance of such strains is a very rare event.

5. Technological Aspects

Cellulases are at present produced only for pharmaceutical and laboratory purposes. The process is a Koji process which is not well-suited to bulk production and consequently the price of the enzyme preparation is high (Table 7). This prevents the use of these cellulases in such processes as treatment of fodder or saccharification of cellulosic materials for SCP- or ethanol-production. The demand for enzymes for these purposes would be very large, provided that an inexpensive enzyme preparation could be produced. In order to be inexpensive enough the process should fulfill the following requirements:
- submerged fermentation, preferably continuous
- short fermentation time
- high enzyme activities
- inexpensive medium.

Table 7. Prices of *Trichoderma* cellulase preparations

Type	Price	Reference
Commercial enzymes:		
Meicellase	$ 30./kg (1971)	56
Onozuka SS P 1500	$ 115./kg (1971)	56
Pancellase	$ 170./kg (1974)	56
Cost estimate:		
Cultivation solution	$.011/l (1975)	65

Research work aimed at developing such a process has reached the pilot plant stage [62, 65]. In all endeavors of this kind the aim has been to produce a suitable cellulase complex using one organism, usually in a single-stage fermentation. The best organism for production of an effective cellulase complex is *Trichoderma viride* [29].
The carbon source in the medium must be a cellulosic material, since enzyme synthesis is repressed by glucose and other rapidly fermentable carbohydrates. This unfortunately leads to slow growth and consequently a long fermentation time. The solution to this problem could be the use of a two-stage process with a growth phase followed by an induction and enzyme production phase [73]. In all experiments so far the induction period needed has been far too long. Another solution would be to use a carbohydrate-limited continuous fermentation. In continuous cultivation on glucose a rather long residence time (ca. 50 hrs) was necessary [64]. One possibility of partly overcoming these difficulties would be the use of a constitutive cellulase-producing mutant. Unfortunately, such a mutant has not yet been produced.

The components of the medium which are most important from an economic point of view are the nitrogen source and the carbon source. Good growth of *T. viride* is obtained on inorganic nitrogen, but good enzyme production is only realized with peptone or some other organic nitrogen source which is usually more expensive. Recycling of the mycelium could partly replace the nitrogen source [65]. This would reduce the production cost by 10%. Some kind of cellulosic waste material can be used as a carbon source, but collecting and pretreatment still precipitate considerable costs.
Pilot plant experiments on the production of *T. viride* cellulases have been carried out by Nystrom and coworkers [62]. They have been able to reduce the lag time substantially and reach maximum cellulase production in 3 to 4 days. The use of a large inoculum (e.g. 10%), pH control and addition of Tween 80 were essential factors.
Based on their pilot scale experiments Nystrom and Allen [65] have designed a production line for concentrated cellulase broths (cf. p. 16). The factory was designed to produce 1000 m^3/d of an enzyme solution containing 1.0 I. U./ml. This enzyme solution is enough to produce 23 to 307 t of sugar. The amount of waste cellulose used as a substrate for this output of sugar is 52 to 429 t.
The cost of the enzyme solution produced was calculated to be $ .011/l (Table 7). The authors estimated that further optimization of the process could lead to a twofold increase in enzyme yield. On the other hand, no cost was assigned to the cellulose contained in the medium. It was assumed that waste material would be used. However, some cost must be assumed for collecting and transporting of this material.
The production of cellulases is at present not economical if the enzymes are intended for bulk processes. Continued research into the improvement of microbial strains and process optimization is still needed in order to enable us to utilize the enormous amounts of cellulosic waste materials available as potential substrates for the fermentation industry.

6. Conclusion

Studies on the production of cellulolytic enzymes and the enzymatic hydrolysis of cellulose have reached the point at which industrial production can be seriously considered. The production of cellulases is complicated by the fact that at least three different types of activity are required for the hydrolysis of native cellulose, endo-β-glucanase, exo-β-glucanase, and β-glucosidase. Furthermore, these enzymes are not single proteins; several isoenzymes differing in properties exist. The mechanism of the enzymatic hydrolysis of cellulose is not yet completely understood. In particular, the initiation of the hydrolysis of fibrous cellulose is obscure. Some recent information indicates that the first step may be oxidative, but this has still to be proved. The idea of selecting different organisms for the production of different enzymes is attractive, but so far the best results have been obtained using one single organism, *Trichoderma viride.* This may result from the fact that the rate-limiting activity in hydrolysis of cellulose is the solubilizing activity. The organisms yielding the best solubilizing activity still produce it at such a level that its concentration is rate-limiting. Hence, the addition of other cellulolytic enzymes does not accelerate the degradation of cellulose. Furthermore, the action of the cellulases is synergistic and not all enzymes from different organisms are necessarily able to act synergistically with each other. Process optimization and genetic improvement of the present strains can still enhance enzyme production. Continuous fermentation would, in principle, be advantageous, but at the present state of knowledge it is not yet feasible.

References

1. Streamer, M., Eriksson, K.-E., Pettersson, B.: Eur. J. Biochem. **59**, 607 (1975).
2. Bucht, B., Eriksson, K.-E.: Arch. Biochim. Biophys. **129**, 416 (1969).
3. Reese, E. T., Siu, R. G. H., Levinson, H. S.: J. Bacteriol. **59**, 485 (1950).
4. Mandels, M., Reese, E. T.: Develop. Ind. Microbiol. **5**, 5 (1964).
5. Wood, T. M., Phillips, R. D.: Nature (London) **222**, 986 (1969).
6. Tomita, Y., Suzuki, H., Nisizawa, K.: J. Ferment. Technol. (Japan) **52**, 233 (1974).
7. Wood, T. M.: in Proc. 4th Intern. Ferment. Symp., Fermentation technology today, G. Terui, Ed., Soc. Ferment. Technol. Osaka 1972, p. 711.
8. Wood, T. M., McCrae, S. I.: Biochem. J. **128**, 1183 (1972).
9. Berghem, L. E. R., Pettersson, L. G.: Eur. J. Biochem. **37**, 21 (1973).
10. Halliwell, G., Griffin, M.: Biochem. J. **135**, 587 (1973).
11. Wood, T. M., McCrae, S. I.: in Symposium on enzymatic hydrolysis of cellulose. M. Bailey, T.-M. Enari, M. Linko, Eds., Aulanko, Finland, 12–14 March 1975. SITRA, Helsinki 1975, p. 231.
12. Pettersson, L. G.: in Symposium on enzymatic hydrolysis of cellulose. Aulanko, Finland, 12–14 March 1975. SITRA, Helsinki 1975, p. 255.
13. Eriksson, K.-E.: in Symposium on enzymatic hydrolysis of cellulose. Aulanko, Finland, 12–14 March 1975. SITRA, Helsinki 1975, p. 263.
14. Wood, T. M.: Biotechnol. Bioeng. Symp. No. 5, 111 (1975).
15. Wood, T. M.: in Proc. 4th Intern. Ferment. Symp., Fermentation technology today, Soc. Ferment. Technol., Osaka 1972, p. 717.
16. Eriksson, K.-E., Pettersson, B.: Eur. J. Biochem. **51**, 193 (1975).
17. Li, L. H., Flora, R. M., King, K. W.: Arch. Biochem. Biophys. **111**, 439 (1965).
18. Cole, F. E., King, K. W.: Biochem. Biophys. Acta **81**, 122 (1964).
19. Wood, T. M.: Biochem. J. **115**, 457 (1969).
20. Wood, T. M.: Biochem. J. **109**, 217 (1968).
21. Eriksson, K.-E.: in Cellulases and Their Applications. R. F. Could, Ed., American Chemical Society Publications, Washington D. C. 1969, p. 58.
22. Eriksson, K.-E., Pettersson, B.: in Proc. 2nd Intern. Biodeterior. Symp., Biodeterioration of materials, vol. 2, Applied Science Pub. Ltd., London 1971, p. 116.
23. Eriksson, K.-E., Pettersson, B., Westermark, U.: FEBS Letters **49**, 282 (1975).
24. Westermark, U., Eriksson, K.-E.: Acta Chem. Scand. **B28**, 204 (1974).
25. Westermark, U., Eriksson, k.-E.: Acta Chem. Scand. **B28**, 209 (1974).
26. Shibata, S., Nisizawa, K.: J. Biochem. **78**, 499 (1975).
27. Goksøyr, J., Eidsa, G., Eriksen, J., Osmundsvag, K.: in Symposium on enzymatic hydrolysis of cellulose. Aulanko, Finland, 12–14 March 1975. SITRA, Helsinki 1975, p. 217.
28. Mandels, M., Weber, J.: Advan. Chem. Ser. **95**, 391 (1969).

29. Mandels, M.: Biotechnol. Bioeng. Symp. No. 5, 81 (1975).
30. Griffin, H. L.: Anal. Biochem. **56**, 621 (1973).
31. Halliwell, G., Riaz, M.: Biochem. J. **116**, 35 (1970).
32. King, K. W.: J. Ferment. Technol. (Japan) **43**, 79 (1965).
33. Leisola, M., Linko, M.: Anal. Biochem. **70**, 592 (1976).
34. Eriksson, K. E., Goodell, E. W.: Can. J. Microbiol. **20**, 371 (1974).
35. Palva, T., Nevalainen, H.: in Proc. 2nd National Meeting on Biophysics and Biotechnology in Finland, A.-L. Kairento, E. Riihimäki, P. Tarkka, Eds., Helsinki 1976, p. 93.
36. Poincelot, R. P., Day, P. R.: Appl. Microbiol. **22**, 875 (1972).
37. Leisola, M., Linko, M., Karvonen, E.: in Symposium on enzymatic hydrolysis of cellulose. Aulanko, Finland, 12–14 March 1975. SITRA, Helsinki 1975, p. 297.
38. Almin, K. E., Eriksson, K.-E., Biochem, Biophys. Acta **139**, 238 and 248 (1967).
39. Child, J. J., Eveleigh, D. E., Sieben, A.: Can. J. Biochem. **51**, 39 (1973).
40. Selby, K., Maitland, C. C.: Biochem. J. **104**, 716 (1971).
41. Mandels, M., Hontz, L., Nystrom, J.: Biotechnol. Bioeng. **16**, 1471 (1974).
42. Halliwell, G.: in Symposium on enzymatic hydrolysis of cellulose. Aulanko, Finland, 12–14 March 1975. SITRA, Helsinki 1975, p. 319.
43. Toyama, N., Ogawa, K.: in Proc. 4th Intern. Ferment. Symp., Fermentation Technology Today, Soc., Ferment. Technol., Osaka 1972, p. 743.
44. Selby, K.: in 1st Intern. Biodeterior. Symp., Biodeterioration of Materials, vol. 1, Applied Science Pub. Ltd., London 1968, p. 62.
45. Boretti, G., Garafano, L., Montecucci, P., Spalla, C.: Arch. Mikrobiol. **92**, 189 (1972).
46. Updegraff, D. M.: Biotechnol. Bioeng. **13**, 77 (1971).
47. Suzuki, H.: in Symposium on enzymatic hydrolysis of cellulose. Aulanko, Finland, 12–14 March 1975. SITRA, Helsinki 1975, p. 155.
48. Enari, T.-M., Markkanen, P., Korhonen, E.: in Symposium on enzymatic hydrolysis of cellulose. Aulanko, Finland, 12–14 March 1975. SITRA, Helsinki 1975, p. 171.
49. Siu, R. G. H.: Microbial Decomposition of Cellulose. Reinhold Pub. Co., New York 1951.
50. Stutzenberger, F.: Appl. Microbiol. **24**, 77 (1972).
51. v. Hofsten, B.: in Symposium on enzymatic hydrolysis of cellulose. Aulanko, Finland, 12–14 March 1975. SITRA, Helsinki 1975, p. 281.
52. v. Hofsten, B., Berg, B., Beskow, S.: Arch. Mikrobiol. **79**, 69 (1971).
53. Srinivasan, V. R.: in Symposium on enzymatic hydrolysis of cellulose. Aulanko, Finland, 12–14 March 1975. SITRA, Helsinki 1975, p. 393.
54. Berg, B., v. Hofsten, B., Pettersson, G.: J. Appl. Bacteriol. **35**, 215 (1972).
55. Wolnak, B.: Present and Future Technological and Commercial Status of Enzymes. National Science Foundation, Rep. No. NSF/RAX/N-73-002, 1972, p. 44.
56. Mandels, M., Sternberg, D., Andreotti, R. E.: in Symposium on enzymatic hydrolysis of cellulose. Aulanko, Finland, 12–14 March 1975. SITRA, Helsinki 1975, p. 81.
57. Reese, E. T., Maguire, A.: Appl. Microbiol. **17**, 242 (1969).
58. Nisizawa, T., Suzuki, H., Nakayama, M., Nisizawa, K.: J. Biochem. **70**, 375 (1971).
59. Nisizawa, T., Suzuki, H., Nisizawa, K.: J. Biochem. **70**, 387 (1971).
60. Mandels, M., Reese, E. T.: J. Bacteriol. **73**, 269 (1957).
61. Nisizawa, T., Suzuki, H., Nisizawa, K.: J. Biochem. **71**, 999 (1972).
62. Nystrom, J. M., Kornuta, K. A.: in Symposium on enzymatic hydrolysis of cellulose. Aulanko, Finland, 12–14 March 1975. SITRA, Helsinki 1975, p. 181.
63. Mandels, M., Reese, E. T.: J. Bacteriol. **79**, 816 (1960).
64. Brown, D. E., Halstead, D. J.: in Symposium on enzymatic hydrolysis of cellulose. Aulanko, Finland, 12–14 March 1975. SITRA, Helsinki 1975, p. 137.
65. Nystrom, J. M., Allen, A. L.: Biotechnol. Bioeng. Symp. No. 6, 55 (1976).
66. Mandels, M., Parrish, F. W., Reese, E. T.: J. Bacteriol. **83**, 400 (1962).
67. Tomita, Y., Suzuki, H., Nisizawa, K.: J. Ferment. Technol. (Japan) **46**, 701 (1968).
68. Toda, S., Suzuki, H., Nisizawa, K.: J. Ferment. Technol. (Japan) **46**, 711 (1968).

69. Ghose, T. K., Pathak, A. N., Bisaria, V. S.: in Symposium on enzymatic hydrolysis of cellulose. Aulanko, Finland, 12–14 March 1975. SITRA, Helsinki 1975, p. 111.
70. Sternberg, D.: Biotechnol. Bioeng. Symp. No. 5, 107 (1975).
71. Mandels, M., Weber, J., Parizek, R.: Appl. Microbiol. **21**, 152 (1971).
72. Morozowa, E. S.: in Symposium on enzymatic hydrolysis of cellulose. Aulanko, Finland, 12–14 March 1975. SITRA, Helsinki 1975, p. 193.
73. Wilke, C. R., Mitra, G.: Biotechnol. Bioeng. Symp. No. 5, 253 (1975).

An Evaluation of Enzymatic Hydrolysis of Cellulosic Materials

M. LINKO
Technical Research Centre of Finland, Biotechnical Laboratory,
Box 192, SF-00121 Helsinki 12, Finland

Contents

Summary

The diminishing one-way resources must be replaced by renewable, plentiful organic materials such as cellulose. Enzymatic hydrolysis of cellulose has been intensively studied in recent years, since acid hydrolysis has not proved to be economically feasible. In spite of the abundance of cellulose, it is not very easy to find suitable cellulosic materials that could be collected from a limited area and would be cheap enough, taking into account collecting, transport, handling, and storage costs. The correct choice of material depends on local conditions. For example, sugarcane bagasse would be useful in certain areas.

An enzyme preparation capable of completely breaking down cellulose is needed for the hydrolysis. *Trichoderma viride* is the most efficient producer of extracellular cellulases known at present. Since several types of cellulases are needed, it is possible that two or more organisms will be used in the future. Pretreatment of cellulosic materials prior to hydrolysis is inevitable, but all the known methods, such as alkali treatment or ball-milling, are rather costly. The product of complete hydrolysis is glucose, which can be used as such for various purposes or as raw material for other products. The economic feasibility of processes based on enzymatic hydrolysis of cellulosic materials is uncertain so far, but the potential of these processes encourages further developmental work.

1. Introduction

Many articles and reviews concerning enzymatic hydrolysis of cellulose begin with statements like "cellulose is the most abundant renewable organic material", or "the vast energy of sun is fixed through photosynthesis on the form of cellulose". Figures exceeding the limits of human comprehension are presented. These statements and figures are, of course, true. The potential of biotechnical processes based on enzymatic hydrolysis of cellulosic materials is enormous. When the limits of utilization of non-renewable resources come closer, cellulose must become a major raw material for food, energy, and other products.

A basic problem in the hydrolysis of cellulose is that, unlike starch, cellulose was not created to act as a carbohydrate reserve which would be readily broken down to glucose whenever necessary. On the contrary, cellulose has been designed to form firm structures responsible for the strength and rigidity of plants. Moreover, the sturdiness of the structures has been secured with an efficient glue, namely lignin.

The natures of cellulose and wood have led to their conventional uses in the manufacture of paper, textiles, and building materials where the strength of the fibers or the rigid structure of wood is essential. In the future these will also, no doubt, remain the main areas of exploitation of the most valuable cellulosic materials. However, there are also huge quantities of cellulosic materials which so far have not been utilized at all. Biotechnical procedures should be developed to exploit "the most abundant organic material in which the vast energy of sun is fixed" in a more efficient way than by burning.

Most cellulosic materials contain three major organic components: cellulose, hemicellulose, and lignin. The economy of processes based on enzymatic breakdown of cellulose does not allow any of these components to be overlooked. Xylan-type hemicellulose typical of hardwood can be used for manufacture of furfural or xylitol; glucomannan-type hemicellulose common in softwood can be hydrolyzed to hexoses, to be used together with glucose obtained from cellulose in, for example, the production of ethanol or single-cell protein; lignin can be utilized for manufacture of certain chemicals–or it can simply be burned.

During the last two years at least three symposia concerning the enzymatic hydrolysis of cellulosic materials have been arranged [1–3]. The proceedings of these symposia comprise a fairly comprehensive survey of the scientific and technical research within this field.

Biological systems for degradation of all plant constituents, including cellulose and lignin, have always existed. Fire has not been the sole tool of nature for recirculation of carbon. However, the existing biological systems are in many cases extremely slow: in certain climatic conditions it might take a hundred years before a stub of pine disappears. This means that the task of developing biotechnical systems for enzymatic hydrolysis of cellulosic materials presents a considerable challenge, bearing in mind that the processes should be economically feasible. At any rate, the potential of such processes is great enough to encourage considerable research activity in the present world of diminishing one-way natural resources.

2. Cellulosic Materials

Cellulose is a constituent of all kinds of plants. Possible raw materials for enzymatic hydrolysis are different wood and non-wood plants or products obtained from these. The forests of the world are generally divided according to the climatic zones, since climate is the most decisive factor affecting the structure of forest [4]. The cool coniferous forest zone includes the vast forests of Siberia and areas in northern Europe and Canada. Long-fiber species such as spruce, pine, fir, and larch are dominating in this zone. The temperate forest zone includes parts of the U. S. A., Central Europe, and the Soviet Union. These forests are rich in hardwood species. They are the most important resources for the wood industries.

The tropical forests on both sides of the equator are extremely heterogenous, with hundreds of species in a small area. The heterogeneity limits the use of these forests. The dry forest zone includes large areas near the equator and parts of southern Europe, India, South America, and Australia. The dry forest area is expanding at the expense of the rain forests as a result of the activities of man. The dry forests have only limited local value.

The world forest resources are summarized in Table 1 [5]. About 22% of the area of the globe is covered by large forests. The area of hardwood forests is somewhat greater than that of softwood forests. The areas richest in forests are the Soviet Union, South America, and North America, all having more than 20% of the total area covered by forests. In Western Europe the respective figure is only 4%.

Table 1. World forest resources available in 1973 [5]. (A typical density of wood is about 0.4 t/m^3)

Region	Estimated volume x $10^9 m^3$		
	Coniferous	Broadleaved	Total
World	103.1	220.4	323.5
North America	26.7	9.4	36.1
Soviet Union	61.5	11.7	73.2
Western Europe	6.1	8.9	15.0
Africa	1.0	33.0	34.0
Asia	1.0	28.0	29.0
Latin America	0.7	123.3	124.0
China	(3.6)	(7.5)	(11.4)
Eastern Europe	2.0	1.5	3.5
Japan	1.0	0.9	1.9
Oceania and South Africa	0.4	1.0	1.4

In addition to the use of wood as fuel, the principal consumers of forest resources are the pulp and paper industries and mechanical industries, which produce sawn timber and plywood. World consumption of paper and board is expected to rise from the 1970 level of 128 million tons to 218 million tons in 1980. This can only be achieved through exploitation of presently unutilized forest resources, including those of many tropical and subtropical regions.

With present harvesting methods, as much as 40% of the organic substance is left in the forests. For this reason harvesting of whole trees and utilization of branches, stumps, and roots is being investigated [4]. Wood-cutting residues, bushes, and small rapidly growing trees are potential raw materials for enzymatic hydrolysis. The main chemical compositions of some woods and pulps are given in Table 2 [6].

Table 2. Chemical composition of some woods and wood pulps [6]

Sample	Yield %	Cellulose %	Gluco-mannan %	Xylan %	Lignin %	Extractives %
Wood						
Spruce	100	41	19	11	27	2
Pine	100	41	18	10	27	4
Birch	100	40	3	33	21	3
Pulp						
Mechanical spruce	98	41	19	11	27	2
Unbleached spruce sulfite	55	79	10	5	5	1
Unbleached pine sulfate	47	77	8	10	5	0.2
Unbleached birch sulfate	54	69	< 1	27	3	< 1
Bleached spruce sulfite	47	85	10	5	+	< 1
Bleached pine sulfate	45	81	8	11	+	0.2
Bleached birch sulfate	50	73	+	27	+	< 1
Neutral sulfite semichemical	80	56	3	26	15	1

Waste paper is also a potential raw material for enzymatic hydrolysis. For example, in the U. S. A., 44.3 million tons of waste paper are formed of which 33.5 million tons are available [7].

However, this material will probably mainly be reused for paper production. The cellulose content of urban waste is quite high. Non-wood plants and fibers include agricultural residues, such as sugarcane bagasse and cereal straw, natural-growing plants such as bamboo, papyrus, and various grasses and non-wood crop fibers such as jute, hemp, manila hemp, sisal, and cotton, which are primarily grown for their fiber content [8].

The availability of some non-wood plant fibrous materials is summarized in Table 3 [4]. The sum total of these materials in the world is very high. The most important are bagasse, bamboo, cotton, and some straws. Their chemical compositions are given in Table 4 [9].

However, there could be one severe drawback in connection with the possible utilization of straws and other agricultural by-products or wastes; the complete removal of all this organic material from the fields may adversely affect the physical and chemical properties of the soil [10]. Similar effects may even occur in forests if the trees are collected down to the last branch and root.

The present crops have not been optimized for production of digestible cellulose. In the future an alternative crop to trees or cane for producing digestible cellulose may be developed.

Table 3. Estimated availability of non-wood plant fibrous materials [4]

Raw material	Potential worldwide availability with present collection methods 1000 metric tons
Sugarcane bagasse	55000
Different straws (wheat, rice, etc.)	88500
Bast fibers (jute, kenaf, etc.)	6099
Leaf fibers (sisal, abaca, etc.)	904
Reeds	30000
Bamboo	30000
Papyrus	5000
Esparto grass	500
Sabai grass	200
Cotton fiber	13500

Table 4. Chemical composition of some non-wood plant materials [9]

Raw material	Ash content %	Lignin %	Pentosans %	α-Cellulose %	Extractives %
Bagasse, fresh	2.4	18.9	30.0	33.4	6.0
Wheat straw	11.0	18.0	28.4	30.5	3.5
Rice straw	17.5	12.5	24.0	32.1	4.6
Bamboo	3.3	20.1	19.6	–	1.2

Animal manure may be a suitable raw material in some cases, for example, in certain areas of the U. S. A. [11]. Some by-products or waste materials of the food industry could also be utilized. Short seasons, often only 1 to 2 months, are problematic. Peat is an interesting material of plant origin. Except as a fuel, it has been poorly utilized. However, the surface layers are quite rich in cellulose and hemicellulose and could be used for enzymatic hydrolysis. Peat resources are enormous, especially in the Soviet Union, and also quite large in some other countries, such as Finland and Canada (Table 5) [12]. Estimated world reserves are more than 220 billion tons.

The chemical composition of peat varies greatly, depending on its origin and age. As an example, the composition of the upper layers of peat deposits has been listed in Table 6 [13]. It has been suggested that most of the essential phenomena of peat formation occur aerobically in whatever comprises the surface layer at any given time; and once this layer is buried beneath fresh peat little further change occurs. The character of peat at any level in a profile thus represents a memory bank of the conditions prevailing when that level constituted the surface. The cellulose content of upper layers is often more than 20% of the organic constituents and the hemicellulose content is also quite high. Even though these values are significantly lower than those of the original plants, the huge peat deposits may be suitable sources for hydrolysis.

Table 5. World distribution of peat [12]

Country	% of world peatland	Total area in million acres
Soviet Union	60.6	181.0
Finland	8.1	24.5
Canada	8.0	23.8
USA	6.0	18.8
Germany	4.2	13.1
Great Britain	4.1	11.0
Sweden	4.0	10.0
Poland	1.2	3.5
Indonesia	1.1	2.8
Norway	0.8	2.5
Ireland	0.7	2.2
Cuba	0.4	1.1
Iceland	0.3	0.8
Japan	0.2	0.5
New Zealand	0.1	0.4
Others	0.1	–

Table 6. Chemical composition of the upper layers of the "Galitskii Moss" peat deposits (High moor peat bed) [13]

Depth cm	Degree of decomposition (Botanical analysis) %	Organic constituents of peat, %				
		Bitumen	Humic acids	Lignin and cutin	Hemi-cellulose	Cellulose
0– 5	0	2.4	1.9	9.2	36.2	21.6
5–24	5	3.5	5.0	10.1	16.6	29.6
25–31	26	7.1	25.4	17.3	13.0	22.6
31–39	15	4.2	15.1	10.5	25.5	21.7
39–44	32	7.4	26.4	21.3	17.2	16.3
44–50	14	4.1	15.4	19.7	15.7	23.8

3. Enzyme Preparations

Production of cellulolytic enzymes has been described by Enari and Markkanen in this volume [14]. The economic feasibility of the hydrolysis of cellulosic materials is essentially dependent on the availability of active cellulase preparations at a reasonable price. There are many cellulolytic organisms that are able to grow on cellulosic materials. However, *Trichoderma viride* is quite outstanding for production of cellulases. The microbiological [15] and technological [16] pioneer work has mainly been carried out at the U.S. Army Natick laboratories. Several other research groups in many countries have recently made valuable contributions in this field.

The enzyme preparation must be able to break down cellulose, probably after some pretreatment, all the way to glucose. At least three types of cellulases are involved in the complete hydrolysis, endo-β-glucanases, exo-β-glucanases and β-glucosidase (cellobiase). At present enzyme preparations produced with *Trichoderma viride* are used alone in laboratory and pilot studies. However, it is not very likely that any one organism would be the best producer of all necessary types of cellulases. Consequently, it can be expected that in the future two or more organisms will be used to produce cellulases, just as several organisms, both bacteria and molds, are utilized for production of different amylolytic enzymes necessary for complete hydrolysis of starch, also a polymere of glucose.

Other organisms studied from the point of view of production of cellulases are, e.g. *Pseudomonas fluorescens* var. *cellulosa* [17] and *Aspergillus awamori* [18]. The cellulase of *Pseudomonas fluorescens* seems to be a constitutive enzyme, the formation of which is controlled by catabolite repression.

The production of cellulases is not necessarily connected with the hydrolysis of cellulosic material, although cellulose is also needed for the enzyme production process. The cultivation solution can be concentrated to 10% of the original volume by ultrafiltration without any loss in activity. Actually, a slight increase in activity has occasionally been noticed [19]. This could be explained by removal of some small-molecular-inhibiting compounds.

Cellulase is commercially available from Japan, but the cost is economically prohibitive for technical use. At present the Japanese production is based on a surface culture method, the so-called Koji-process [20]. Nevertheless, at the present level of development of the relevant process technology, the best strains of *Trichoderma viride* should even now enable us to produce cellulases at a relatively low price in full-scale submerged culture.

4. Hydrolysis of Cellulose

4.1 Pretreatment

Native cellulose is very resistant to enzymatic hydrolysis. The highly crystalline structure and the presence of lignin effectively prevent the attack of cellulases, making the hydrolysis slow and incomplete. For a practical hydrolysis process it is necessary to treat the cellulosic material in some way prior to the use of enzymes.

Cellulose fibers are of similar structure, irrespective of their function in the plant. For example, the cellulose framework of wood fibers is essentially the same as that of cotton seed hair, although the wood fibers form the supporting tissue of the plant and the cotton seed hair serves to spread the seeds of the plant as widely as possible [21]. The main divergences between various types of fibers arise from the different dimensions of the fibers and, in particular, from the differences in the nature of other substances present and their location in the fiber. For example, cotton is almost pure cellulose; its cellulose content is more than 90% of the dry matter, whereas wood cells contain 40 to 45% cellulose [22, 23] and cereal straws about 30% [24].

The problems in the hydrolysis of native cellulose are partly geometric; there is actually no space for the enzyme between the cell wall structures [25]. The goal for pretreatment of cellulosic materials is the exact opposite of that of the cellulose industry, in which the fibers are required to be as untouched as possible. Therefore, all the experience of the cellulose manufactures should be turned upside down in order to find efficient pretreatment methods for hydrolysis of cellulosic materials. On the other hand, some sulfite processes lead to complete reactivity of the cellulose produced [26]. Consequently, some procedures resembling the manufacture of cellulose could perhaps be modified to act as pretreatments for hydrolysis.

The pretreatments are aimed at loosening the highly crystalline structure of cellulose and extending the amorphous areas. The removal of lignin is also essential. A decrease of one third in the lignin content of hardwood or two thirds in that of softwood increases the digestibility of these materials to 60%, which is equivalent to hay for ruminants [27, 28]. Partial delignification can be achieved by using Kraft reagent [28], sodium hydroxide, alone or followed by peracetic acid [29–31], sodium hypochlorite in acetic acid and chlorine dioxide gas [27]. Treatment with sulfur dioxide is equivalent to the conventional ammonium bisulfite process for production of cellulose and can be used to break the binding of lignin to carbohydrates without any actual removal of lignin [32]. Biological delignification is an interesting possibility [33–36]. It has been found that the treatment of wood with white rot fungi capable of breaking down lignin leads to a more efficient hydrolysis by cellulases or rumen fluid [37].

Treatments with alkali are efficient, but not inexpensive. Some alkali treatments reported are listed in Table 7. Sodium hydroxide swells the cellulose fibers and causes depolymerization. The lowest concentration of sodium hydroxide that causes intramicellar swelling is 8%. At lower concentrations intermicellar swelling can be observed [46]. A short treatment with 0.1 to 0.5% sodium hydroxide causes esterification of uronic acids and saponification of acetates, which leads to swelling and loosening of the structure. If the treatment time is longer, some lignin and hemicellulose are dissolved, causing dry-matter loss. For example, after six hours in 1% sodium hydroxide at 24° C, 9% of wood is dissolved and after three hours in 4% sodium hydroxide at 60° C, 17%, respectively. Zinc oxide added to the alkaline solution enhances the swelling of cellulose [47, 48].

Table 7. Alkali pretreatment methods

Material	Dry matter g/ml	NaOH %	Temperature ° C	Time min	Reference
Aspen, cotton	0.05	1	30	60	38
Newsprint	0.02	2	70	90	39
Bagasse		2	72	60	39
Cotton, spruce, beech, oak	0.05–0.1	0.5–17.5	20	30	40
Rice straw		2–4	100–160	10–60	41
Bagasse, rice straw	0.14	1	100	180	29
Bagasse	0.04	7	80	>180	28
Newsprint, rice straw	0.03	1	22	960	42
Bagasse		0.4	120	15	43
Barley straw	0.06	40	(high pressure)		44, 45

The alkali treatment has also been criticized [39, 49]. Swelling in alkaline solutions enhances the enzymatic hydrolysis, but suspensions of 4 to 5% are already too thick for pumping and mixing. Loss of material is a severe disadvantage. Furthermore, heating of xylose in an alkaline solution may cause formation of compounds inhibiting the growth of microorganisms. This is a serious drawback if the ultimate goal is production of biomass [50].

Steam treatment at 160 to 170° C breaks acetyl groups, leading to enhanced hydrolysis. The digestibility of straw, and especially hardwood, is significantly improved by steam treatment [27]. Various chemical treatments are efficient, but there are also limitations in the use of strong chemicals: the treatment is expensive; the material has to be washed and neutralized, causing material loss and giving rise to problems associated with waste outflow and the chemical may be toxic.

Inorganic acids cause swelling and solubilization of cellulose. The acids may also react with cellulose by forming esters and hydrolyzing glycosidic bonds, but through use of the correct concentration of acid these reactions can be eliminated [39]. For example, maximum swelling of cotton fibers is achieved with 62.5 to 65.5% sulfuric acid. Phosphoric acid depolymerizes cellulose less than do other mineral acids. Therefore it is commonly used for swelling of cellulose without any significant hydrolysis [23, 39, 47, 51, 52].

Several physical pretreatment methods have also been studied. Milling to a very fine powder is one of the most efficient pretreatments tested. An important advantage is the possibility of using relatively concentrated suspensions, which means that high sugar concentrations are possible in the hydrolyzates [53]. Ball-milling gives the best results, but the cost is high [39]. Ball-milling after alkali treatment is even more efficient than alkali treatment alone. Alkali treatment of bagasse seems to be more efficient than ball-milling.

Radiation breaks polysaccharide molecules. However, high doses are evidently too expensive for industrial application [27]. Heat treatment is usually combined with some mechanical or chemical pretreatment. The order of treatments affects the results [41, 49, 51, 54–56]. Heating alone leads to a decrease in porosity and retardation of hydrolysis [57].

Summing up the experience with pretreatments: There are efficient methods such as alkali treatment and ball-milling, but their cost still makes the pretreatment one of the most critical phases in terms of the overall economics.

An ideal case would comprise that situation where pretreatment does not precipitate any extra cost at all. This is possible if the material has already been used for some process prior to the hydrolysis of cellulose, for example, manufacture of furfural or xylose [58]. If the use of hemicellulose for manufacture of furfural is economical *per se*, this would essentially eliminate the cost of pretreatment of cellulose. The conditions of high temperature (176° C), acid environment, and sudden release of pressure ensure that the manufacture of furfural includes an exceptionally efficient pretreatment. On the other hand, even if the furfural production process is economically feasible, the market for furfural is limited. It can be thought of as a plausible first stage in a few special cases only. The manufacture of xylose for production of xylitol is less efficient as a pretreatment phase for hydrolysis of cellulose [59].

A natural pretreatment is also achieved during the formation and decomposition of peat. The slow natural decomposition of peat makes the cellulose and hemicellulose relatively susceptible to enzymatic hydrolysis without any pretreatments. Here the economic optimum is a compromise between two opposing factors; decomposition of peat makes the hydrolysis essentially easier, but also leads to loss of cellulose and hemicellulose. Some typical examples of the composition and susceptibility of Finnish peats to hydrolysis are summarized in Table 8. Steam treatment in an autoclave significantly improves the results of hydrolysis [59].

Table 8. Enzymatic hydrolysis of some peat samples with different degrees of decomposition [59]. Hydrolysis at pH 5.0, 45° C for 48 h with *Trichoderma viride* cellulase (c), commercial hemicellulase, Société Rapidase, Seclin, France (hc), and a mixture of cellulase and hemicellulase (c + hc)

Degree of decomposition	Holocellulose %	Result of hydrolysis, % reducing sugars/holocellulose		
		c	hc	c + hc
1	77.7	4	2	
2	56.5	15	5	24
3	40.0	27	7	34
5–6	34.8	26	19	37
5–10	15.2	34	20	59
10	10.7	25	27	64
5–10	9.7	52	52	103

4.2 Hydrolysis

Factors affecting the hydrolysis of cellulosic materials include: type of substrate, pretreatment, characteristics of the enzyme preparation(s), temperature, time, pH, substrate concentration, reuse of enzyme, and type of reactor. Many of these factors are interdependent, making the whole process quite complicated. Bearing in mind the stringent demands of economic feasibility, there are many limitations on the process design and many of these limitations are mutually incompatible. For example, the cost of enzyme does not allow high enzyme concentrations, but slow reaction caused by low enzyme concentration does not bear out the cost of investment in the process. Similarly, the reuse of enzyme does not permit a high reaction temperature, but the reaction is too slow at low temperatures.

In an ideal case the reaction would be rapid and complete, leading to a high glucose concentration without any substantial loss of enzyme, which could therefore be reused. At present, the hydrolysis of cellulose does not approach this ideal situation, but the results achieved are at least promising enough to encourage further development and optimization of the process.

On a small scale hydrolyzates with 30% of glucose have been achieved using high enzyme- and substrate concentrations and a long reaction time [53]. The most economical substrate concentration is near 10% [29–31].

Adsorption or ultrafiltrations has been proposed for reuse of enzyme [60]. Cellulose adsorbs cellulase efficiently at pH 4 to 5 and at a temperature of 25 to 50° C, i.e. at the optimal range for enzyme activity. Adsorption is rapid at the beginning of hydrolysis. During hydrolysis the enzyme is liberated and bound again to new substrate. Adsorption depends on the concentration and particle size of cellulose and on the enzyme component [61]. Reports on the desorption of cellulase are contradictory [61–65]. Ultrafiltration for reuse of enzyme has been studied only with pure cellulose as the substrate. In this case no problems with lignin and other compounds remaining in the reactor are encountered.

An increase in temperature from 45 to 55° C enhances the hydrolysis, but impairs the possibilities of reusing the enzyme [65]. During the first hours the rate of hydrolysis is higher at 55° C, but after 10 h 45° C is advantageous. At a temperature of 45° C the usual reaction time has been longer than one day, for example 40 h. Efficient mixing accelerates the hydrolysis, but may also inactivate the enzyme through the shear stress effect. High substrate concentrations cause technical problems in mixing [59]. The pH optimum is slightly lower than 5.0, but higher than 4.0. Since the pH optimum range is relatively narrow, pH control during hydrolysis may be necessary. The pH tends to decrease during the hydrolysis [58]. When Solka Floc is used as a substrate, two distinct pH optima can be seen, one at 4.25 and another at 4.75 (Fig. 1). The situation is similar when using waste from the furfural manufacturing process as a substrate, although the optimum range is sharper and the lower optimum less clear (Fig. 2). The

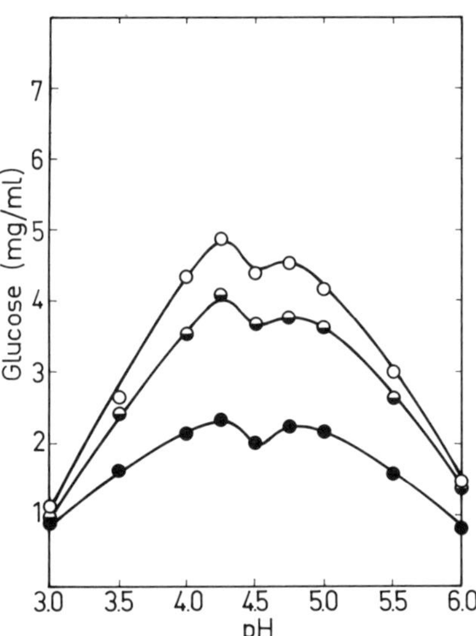

Fig. 1. Effect of pH on the activity of *Trichoderma viride* cellulases. Formation of glucose with a dilute enzyme solution from a 2% suspension of Solka Floc within 1 (●—●), 2 (◒—◒), and 3 (○—○) days of hydrolysis [58]

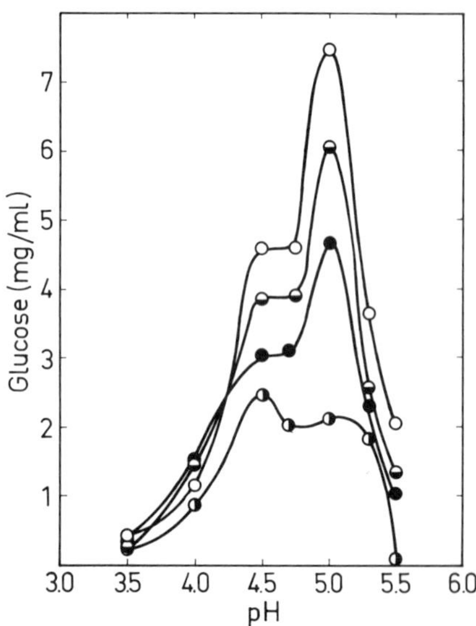

Fig. 2. Effect of pH on the activity of *Trichoderma viride* cellulases. Formation of glucose with a dilute enzyme solution from a 4.6% suspension of furfural process waste within 1 (◑—◑), 2 (●—●), 3 (◒—◒) and 4 (○—○) days of hydrolysis [58]

fact that there are two optima may result from the presence of different components in the cellulase complex.
The product of hydrolysis, glucose, is inhibitory even at low concentrations [30]. It would be advantageous to remove glucose continuously from the hydrolysis mixture to eliminate the inhibitory effect. The strong inhibitory effect of lactose should also be mentioned, since commercial Japanese cellulase preparations contain lactose.
With *Trichoderma viride* cellulase it has been possible to hydrolyze a 10% suspension of waste from the furfural process at pH 5 and 45° C almost completely in 20 hrs [59]. The concentration of reducing sugars in the hydrolyzate was 5%. The furfural process waste contained 54% holocellulose.

4.3 Reuse of Enzyme

Because of the high cost of enzyme, enzyme recovery has been studied. It is not yet clear whether enzyme recovery will be by ultrafiltration of by an adsorption train [60]. Enzyme recovery by adsorption has been suggested as follows [65]. Ground cellulosic waste material is contacted countercurrently for enzyme recovery in five mixer-filter stages with filtrate solution from the hydrolysis section. Following hydrolysis the effluents are filtered in a vacuum drum filter to separate out the liquid and solid phases. After filtration the liquid stream enters the adsorption train to recover the enzyme in the liquid phase. The cellulosic solids with the recovered enzyme protein adsorbed on them are fed to the hydrolysis vessel. The filtered solids from the hydrolysis enter a wash train for recovery of the adsorbed enzyme by washing with process water in six countercurrent stages. The liquid stream is fed to the hydrolyzers and the spent solids are used as a source of fuel.
However, the desorption technique seems to be unclear. The adsorbed enzyme may even be irreversibly denatured. Therefore, at present it has to be assumed that no enzyme recovery from the hydrolyzed solids is possible. In practice, not more than 50% enzyme recovery could be expected from the adsorption systems. Even this is optimistic.
Some recent results indicate that after an actual hydrolysis of waste from the furfural process lasting two days the activities left in the hydrolysate are low and variable. An average of eight runs yielded: activity toward dyed Avicel cellulose, 14%; activity toward carboxymethyl cellulose, 35%; β-glucosidase, 13%; and the so-called filter paper activity, 0% of the original [59]. Because of the rapid and effective adsorption of the cellulases on the substrate it is difficult to measure the loss of activity during the course of hydrolysis. In any case, the residual activities may be so low that it is questionable whether arrangements for reuse of enzyme would be feasible. If reuse is not feasible, the optimum temperature may be relatively high, perhaps 55° C.

5. Final Products

The product of complete hydrolysis of cellulose is, of course, glucose. Another possible product is cellobiose. The composition of the syrups depends on many factors, for example, the relative β-glucosidase activity of the cellulase preparation used.

Cellobiose is reported to have 75% of the sweetening power of glucose and it is non-metabolizable in the human digestive tract. If it should prove that the lower intestinal microflora do not metabolize cellobiose either, then it could be a useful dietetic sweetener. Cellobiose can be used as a controlled glucose-releasing agent in the baking industry. However, the total market for cellobiose will probably be quite limited.
The main hydrolysis product, glucose, is a very useful sugar. It can be used as such for various purposes or as raw material for many other products such as fructose, ethanol, single-cell protein, or almost any fermentation product. The final product to some extent governs the details of the total procedure and perhaps also the choice of raw material. For example, the hydrolyzate may contain impurities which disturb or inhibit fermentation processes, such as SCP-production, or make the purification of glucose difficult. At least in this respect the enzymatic hydrolysis is superior to acid hydrolysis, which tends to produce many compounds toxic to microorganisms, e.g. furfural. A process in which enzyme production, hydrolysis of cellulose, and SCP-production are separated gives a free choice of SCP-organism, in contrast with the direct cultivation of cellulolytic organisms on cellulosic substrates, such as the interesting process developed by the Louisiana State University group [66].
Cheap glucose from waste cellulose would, no doubt, find markets in many fields. Ten to 20% alcohol can be used to extend gasoline without engine changes and with improved operating efficiency [67]. It is possible that the production of industrial ethanol will turn from the synthetic process based on ethylene back to fermentation. However, it should be kept in mind that these uses of glucose from waste cellulose are possible only if the glucose produced is really cheap. In most cases it must at least compete with glucose produced from cereal starch.

6. Process Examples

A simplified scheme of some basic alternatives for processes and products based on the hydrolysis of cellulose is illustrated in Fig. 3 [68]. When starting from a cellulosic waste material the two other main components, hemicellulose and lignin, are also present, although they are not included in the scheme. However, they should also be utilized in some way. Hemicellulose could be used for production of furfural or xylitol prior to the hydrolysis of cellulose or for production of pentoses and hexoses simulaneously with the hydrolysis of cellulose. Lignin could at least be utilized for production of some energy by burning, perhaps also for production of some chemicals or glue. A concept for versatile utilization of cellulose by biological means is delineated in Fig. 4 [69]. This scheme indicates the flexibility of an approach based on enzymatic digestion of pretreated cellulosic materials yielding predominantly glucose.
As an alternative, acid hydrolysis of cellulosic materials should also be kept in mind. Some essential features for comparison have been compiled in Table 9. The most important advantage of acid hydrolysis is perhaps reaction rate; the time needed is typically 20 min, whereas enzymatic hydrolysis takes several hours. The severe corrosion causes technical difficulties and high investment costs in acid hydrolysis. On the other hand,

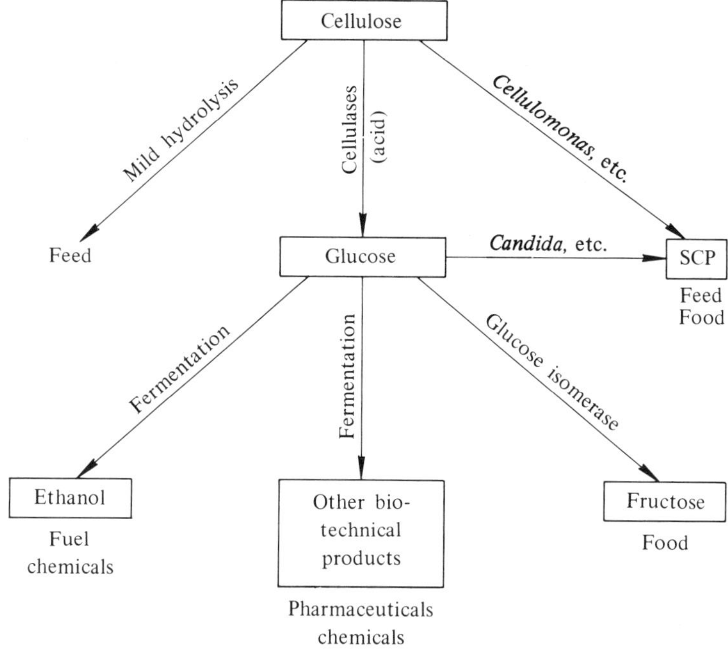

Fig. 3. A simplified scheme of some processes and products based on hydrolysis of cellulose [68]

Table 9. A comparison of some features of acid hydrolysis and enzymatic hydrolysis of cellulosic materials [68]

	Acid hydrolysis	Enzymatic hydrolysis
Pretreatment	may be necessary	necessary
Rate of hydrolysis	fast (minutes)	slow (hours)
Temperature	high (e.g. 200° C)	low (e.g. 45° C)
Pressure	overpressure	atmospheric pressure
Yield	varies depending on material and process details	varies depending on material and process details
Formation of interfering by-products	probably formed	not likely
Industrial processes in use	yes (in Soviet Union)	no (pilot plant only)
Economical feasibility	?	?

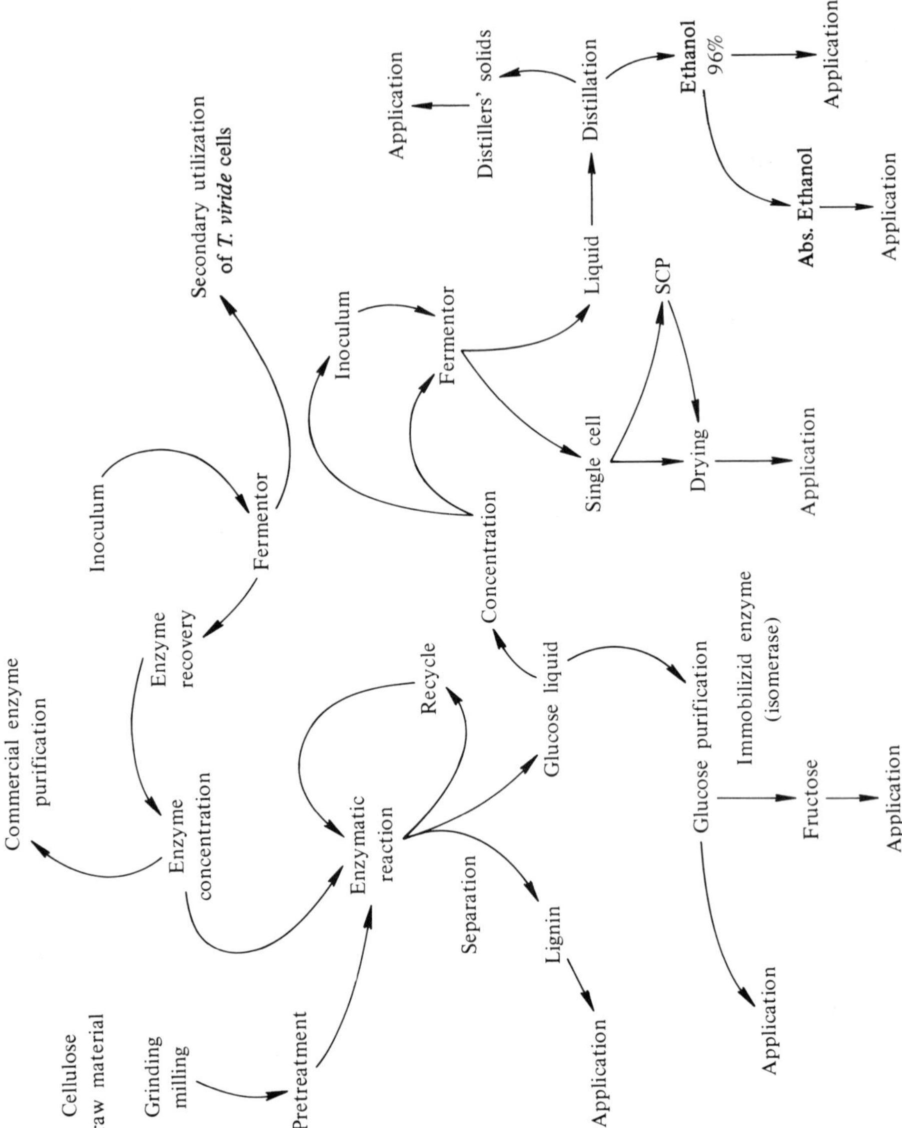

Fig. 4. Concept of versatile utilization of cellulose by biological means [69]

the long period of time needed for enzymatic hydrolysis means a bulky plant, but simple constructions and non-expensive materials can be used, since the temperature is low, the pressure atmospheric, and the pH relatively near neutral.

One serious drawback of acid hydrolysis may be the formation of interfering compounds, for example, decomposition products from sugars. Even a small quantity of compounds toxic to microbes may seriously inhibit ethanol fermentation of SCP-production. These difficulties are not likely to be encountered in processes based on enzymatic hydrolysis because of the selectivity of the enzymatic reactions. Nevertheless, even enzymatic hydrolysis does not lead to pure glucose solutions, since the raw material is not pure cellulose and the hydrolysis is incomplete.

Acid hydrolysis of cellulosic materials is practised industrially in the Soviet Union, but calculations of economical feasibility have not been presented. Many figures for the feasibility of the enzymatic hydrolysis have been suggested, both optimistic and pessimistic, but they are based on laboratory or pilot studies–and presumptions.

The flow diagram for a supposed process including production of cellulases and enzymatic hydrolysis of cellulosic materials is outlined in Fig. 5 [65]. The production of cellulases need not necessarily be combined with their use. On the other hand, the sugar solution produced in the hydrolysis could advantageously be used directly for production of ethanol or SCP, for example, because the sugar concentration may be suitable as such. Besides the conventionally used *Candida* species *Paecilomyces varioti* is one of the suitable organisms for SCP-production. It is cultivated industrially on sulfite waste liquor [70].

7. Economic Feasibility

In some calculations the cost of raw material has been neglected; even a negative value has been suggested because of the possible pollution problems caused by the cellulosic waste in question. However, the raw material cost will most probably not be very low, since the costs for collection, handling, transport, and storage are relatively high. In most cases there are also alternative uses for the material, at least burning for energy. It is realistic to presume a cost of about 10 cents per kilogram for the material. This may be as much as one half of the total cost of the hydrolysis process [60]. Some prices reported for cellulosic materials are presented in Table 10. The price of cellulase preparations is difficult to predict, since the markets have been very limited [73]. The present status of the technology would perhaps enable us to produce cellulases industrially at a cost of about U.S. $ .011 per liter of cultivation solution, the activity of which would be 1000 IU per liter [74]. Depending on the conditions for hydrolysis, this would imply a cost of U.S. $ .037 to .485 per kilogram of sugar produced. The cost of enzyme could be decreased by further improvement of the *Trichoderma viride* strain and development of the production technology. The use and recirculation of enzyme could also be optimized.

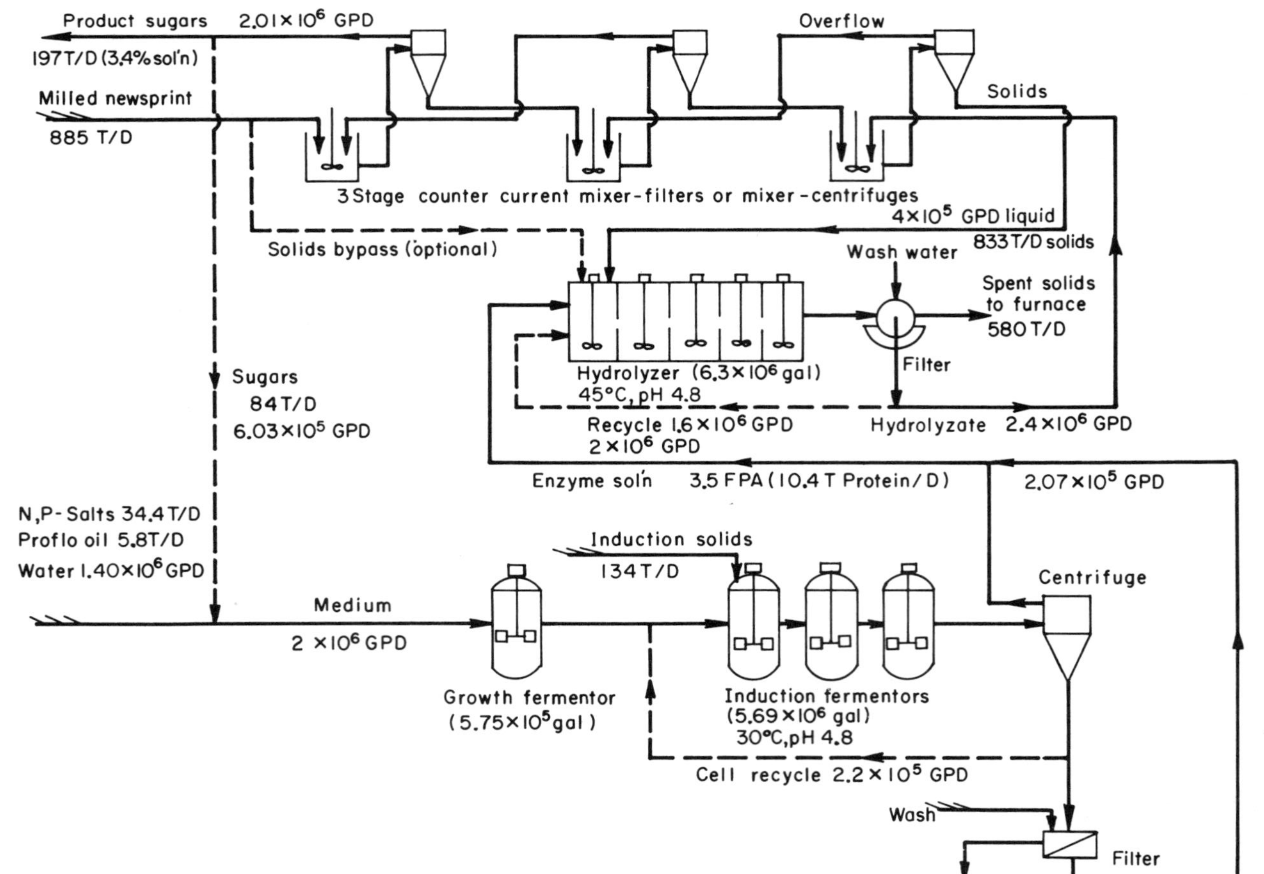

Fig. 5. Flow diagram for a hydrolysis process [65]

Table 10. Costs of some cellulose sources

Material	Cost		Reference
	$ / ton	$ / ton of cellulose	
Pulp, sulfite or sulfate, bleached	325–340	368–395	71
Pulp, sulfate unbleached	310–315	348–366	71
Sugarcane bagasse	15– 17	34– 37	71
Hay, clover and grass	40– 55	126–551	71
Newsprint (waste)	14	24	71
Corrugated (waste)	10	14– 15	71
Urban waste		10– 30	72

However, it is too optimistic to predict any essential changes in the enzyme cost merely as a result of some minor improvements in the details of the processes, since quite thorough studies have already been carried out. A real break-through in cutting the enzyme costs may be achieved, if new organisms, perhaps two or more organisms efficiently producing the various components of the cellulase complex, were to be found for the production of cellulases. This line of development also calls for detailed biochemical knowledge. If relatively efficient cellulase-producing bacteria are found, the genetic manipulation becomes easier. Constitutive organisms have, of course, special advantages.

The pretreatment cost may be very decisive with regard to the overall economics. Both investment and operational costs of physical treatments are high, but no harmful wastes are formed. Investment costs for chemical treatments are lower, but the necessary chemicals and the disposal of wastes may precipitate high costs.

The cost of physical pretreatments may be about U.S. $ .2 per kilogram [75]. If the cost is $ .15 to .18/kg, the additional price for the sugar produced is $ .22 to .44/kg [72]. According to Nystrom and Allen [74] the cost for ball-milling is about $ .20 per kilogram sugar produced.

Pretreatment is evidently always necessary. However, as mentioned above, there may be special cases in which there is no cost for pretreatment. This is possible if the material has been used for another process in such a way that the necessary modification of the fibers is achieved, for example, for production of furfural, prior to the hydrolysis of cellulose.

Wilke [76] has presented calculations based on the following process details. Feeding capacity is 885 tons newsprint per day (Wall Street Journal, lignin content 23%) and a 50% hydrolysis is achieved in 40 h at 45° C. The enzyme activity is 3.5 FPA (= filter paper activity) and the enzyme reuse 34%. The production of glucose is 238 tons per day as a 4% solution. According to Wilke the production cost of sugar is then $ .11/kg. The degree of hydrolysis greatly affects the cost; with an 80% hydrolysis the cost is only $ .07/kg. The cost of the raw material is not included in these figures. If the value of waste paper is $ 20/t, the additional cost for production of sugar is $ .08/kg. Even more significant is the possible cost of pretreatment, which may be as high as all the other costs combined. According to Brandt [72] the minimum cost for glucose produced by enzymatic hydrolysis of waste cellulose is almost \$ .50/kg. The current world market price for raw sugar is only about \$ .30/kg.

The two other main components of most cellulosic materials, hemicellulose and lignin, have not usually been included in the total cost analysis. However, their role is quite significant. Mere burning of the lignin could bring back about half of the raw material costs, not to speak of the potential of more sophisticated ways of using it.

A comparison between hydrolysis of maize starch and cellulose is represented in Table 11 [77]. This comparison is not too unfavorable for cellulosic materials. It should be pointed out that the sugar stream is not pure glucose solution, either from maize starch or cellulosic materials. Even small amounts of impurities may be highly disadvantageous because of inhibition of fermentation or difficulties in crystallization–for example, if pure glucose is the final product.

Table 11. Comparison between maize and cellulosic materials used for glucose production [77]

	Maize		Cellulose	
Cost	Starch	20 c/kg	Cellulose	4–11 c/kg
	Enzymes and processing	4 c/kg	Enzymes and processing	15–22 c/kg
	Total	24 c/kg	Total	19–33 c/kg
Feed stream	Amylose		Cellulose	
	Amylopectin		Hemicellulose	
	Some protein		Lignin	
Sugar stream	Concentration 30%		Concentration 5–10%	
	Besides glucose		Besides glucose	
	5% trimers, maltose, isomaltose		5% other compounds (?)	

The maximum permissible price for glucose to be used for production of ethanol is about \$ .08/kg [60, 78, 79]. This figure is very low and indicates that only low-value wastes could be used for enzymatic hydrolysis of cellulosic materials. The situation is similar for SCP-production [60]. Isomerized glucose syrup for food applications could tolerate a significantly higher glucose price and thus relatively pure cellulose could be used as starting material [60].

An important application of cellulases will probably be treatment of cellulosic feed materials to improve their digestibility. However, details of such treatments and their costs have not been presented.

There are also several factors related to economics which cannot easily be given in plain figures. These factors include the balance of payments within a country, regional goals, such as full employment and use of local resources, long term environmental economy, and national or international crises. These are, no doubt, important points that may give strong motivation for research or even the construction of an industrial plant.

Actually there are still so many uncertain features in the hydrolysis of cellulosic materials that the calculations should not be regarded as much more than an educated guess. For further development it would now be vital to build a relatively large pilot plant, preferably including all the phases from production of cellulases to the use of glucose produced through hydrolysis. Only then could reliable data be obtained for the feasibility studies. This would also reveal all the areas in need of improvement in order to make the processes able to compete with those based, for example, on starchy materials.

8. Conclusions

There is really no lack of cellulose in the world. However, this does not necessarily mean that suitable raw materials for enzymatic hydrolysis can easily be found. The long and strong fibers of the most valuable woody materials will in the future also be used mainly for pulp and paper production. Even waste paper will in the first place be recirculated to paper mills. The material must be cheap, including costs for collection, transport, handling, and storage. Even a very cheap material is not necessarily cheap at all when it must be collected and brought to the mill. The selection of a suitable material poses an essential problem. The proper choice depends on local conditions.

An efficient, low-cost enzyme preparation is needed for the hydrolysis. *Trichoderma viride* is the best producer of extracellular cellulases known so far. It is not easy to further develop its ability to produce these enzymes. A constitutive mutant would bring about substantial improvement. Perhaps two or more organisms can be found for production of the different components of the cellulase complex. This calls for further basic biochemical and microbiological work.

A suitable pretreatment method is an indispensable requirement for the hydrolysis of cellulosic materials. Unfortunately, efficient pretreatments, chemical or physical, are quite costly. More technical research should be carried out in this field. An ideal case in which an efficient pretreatment is achieved simultaneously with a process such as the production of furfural must be seen as exceptional.

Suitable conditions for hydrolysis are relatively well known. However, here again, more biochemical knowledge of the different cellulases would be valuable. The reuse of enzyme is economically very significant, since the price of enzyme is high. Unfortunately, less than 50% enzyme recovery can be expected from adsorption systems, maybe even much less. The desorption technique seems to be unsolved.

Complete hydrolysis of cellulose produces glucose. The final product could be ethanol, SCP, or almost any other fermentation product, because glucose is suitable as a source of carbon and energy for a number of processes. The overall economics would perhaps be on a sounder basis if the final product were more expensive than ethanol or SCP. Isomerization of glucose to fructose is an interesting possibility. An important application of cellulases will probably be treatment of cellulosic feed materials to improve their digestibility.

The overall economical feasibility is very much dependent on efficient use of all main components of the plant materials, cellulose, hemicellulose, and lignin. The economy is not great enough to allow any of these to be disregarded. The present status of feasibility studies is not much above the educated-guess level. The best, and actually the only, way to resolve the many uncertain points is to build a plant—and it should not be too small. Considering all the important features of the utilization of cellulosic materials, an enormous and renewable resource, this should be done.

References

1. Wilke, C. R., Ed.: Cellulose as a chemical energy resource. Biotechnol. Bioeng. Symp. No. 5, John Wiley & Sons, New York 1975, p. 361.
2. Bailey, M., Enari, T.-M., Linko, M., Eds.: Symposium on enzymatic hydrolysis of cellulose. Aulanko, Finland, 12–14 March 1975. SITRA, Helsinki 1975, p. 525.
3. Gaden, E. L., Mandels, M. H., Reese, E. T., Spano, L. A., Eds.: Enzymatic conversion of cellulosic materials: technology and applications. Biotechnol. Bioeng. Symp. No. 6, John Wiley & Sons, New York 1976, p. 319.
4. Virkola, N-E.: In Symposium on enzymatic hydrolysis of cellulose. Aulanko, Finland, 12–14 March 1975. SITRA, Helsinki 1975, p. 23.
5. Pringle, S. L.: EUCEPA International Symposium, New forest resources for the paper industry and their application. Madrid, May 6–8 1974, reprint No. 13.
6. Rydholm, S.: Pulping Processes. John Wiley & Sons, New York 1965, p. 1269.
7. Holloway, R.: Paper presented in Symposium on enzymatic conversion of cellulosic materials: technology and applications. Newton, Mass., September 8–10, 1975.
8. Atchison, J. E.: EUCEPA International Symposium, New forest resources for the paper industry and their application. Madrid, May 6–8 1974, reprint No. 25.
9. Clark, T. F.: In Pulp and paper manufacture, Vol. 2. Control secondary fiber structural board coating, 2nd ed., R. G. Macdonald & J. N. Franklin, Eds., McGraw-Hill, New York 1969, p. 1.
10. Sloneker, J. H.: In Symposium on enzymatic conversion of cellulosic materials: technology and applications. Biotechnol. Bioeng. Symp. No. 6, John Wiley & Sons, New York 1976, p. 235.
11. Griffin, H. L., Kaneshiro, T., Kelson, B. F., Sloneker, J. H.: In Symposium on enzymatic hydrolysis of cellulose. Aulanko, Finland, 12–14 March 1975. SITRA, Helsinki 1975, p. 419.
12. Fogarty, W. M., Griffin, P. J., Ward, J. A.: Technol. Ireland 1973 April 4, p. 21.
13. Kurbatov, I. M.: In 2nd Intern. Peat Congr., Leningrad 1963, Translations, Vol. 1. Her Majesty's Stationery Office, Edinbourgh 1968, p. 133.
14. Enari, T.-M., Markkanen, P.: In Advances in Biochemical Engineering, Vol. 5. Springer-Verlag, Heidelberg 1977, p. 1.
15. Mandels, M., Weber, J.: Advan. Chem. Ser. **95**, 391 (1969).
16. Nystrom, J. M., Kornuta, K. A.: In Symposium on enzymatic hydrolysis of cellulose. Aulanko, Finland, 12–14 March 1975. SITRA, Helsinki 1975, p. 181.
17. Suzuki, H.: In Symposium on enzymatic hydrolysis of cellulose. Aulanko, Finland, 12–14 March 1975. SITRA, Helsinki 1975, p. 155.
18. Enari, T-M., Markkanen, P., Korhonen, E.: In Symposium on enzymatic hydrolysis of cellulose. Aulanko, Finland, 12–14 March 1975. SITRA, Helsinki 1975, p. 171.
19. Nystrom, J. M.: personal communication.
20. Toyama, N.: In Symposium on enzymatic conversion of cellulosic materials: technology and applications. Biotechnol. Bioeng. Symp. No. 6, John Wiley & Sons, New York 1976, p. 207.
21. Sihtola, H., Neimo, L.: In Symposium on enzymatic hydrolysis of cellulose. Aulanko, Finland, 12–14 March 1975. SITRA, Helsinki 1975, p. 9.
22. Cowling, E. B.: In Advances in Enzymatic Hydrolysis of Cellulose and Related Materials. E. T. Reese, Ed., Pergamon Press, London 1963, p. 1.

23. Cowling, E. B., Brown, W.: Advan. Chem. Ser. **95**, 152 (1969).
24. Salo, M.-L.: Acta Agralia Fennica **105**, 8 (1965).
25. Cowling, E. B., Kirk, T. K.: In Symposium on enzymatic conversion of cellulosic materials: technology and applications. Biotechnol. Bioeng. Symp. No. 6, John Wiley & Sons, New York 1976, p. 95.
26. Andren, R. K., Erickson, R. J., Medeiros, J. E.: In Symposium on enzymatic conversion of cellulosic materials: technology and applications. Biotechnol. Bioeng. Symp. No. 6, John Wiley & Sons, New York 1976, p. 177.
27. Millett, M. A., Baker, A. J., Satter, L. D.: In Cellulose as a chemical and energy resource. Biotechnol. Bioeng. Symp. No. 5, John Wiley & Sons, New York 1975, p. 193.
28. Dekker, R. F. H., Richards, G. N.: J. Sci. Food Agric. **24**, 375 (1973).
29. Toyama, N., Ogawa, K.: In Proc. IV Intern. Ferment. Symp., Fermentation technology today. G. Terui, Ed., Soc. Ferment. Technol., Osaka 1972, p. 743.
30. Toyama, N., Ogawa, K.: In Symposium on enzymatic hydrolysis of cellulose. Aulanko, Finland, 12–14 March 1975. SITRA, Helsinki 1975, p. 375.
31. Toyama, N., Ogawa, K.: In Cellulose as a chemical and energy resource. Biotechnol. Bioeng. Symp. No. 5, John Wiley & Sons, New York 1975, p. 225.
32. Clarke, S. D., Dyer, I. A.: J. Anim. Sci. **37**, 1022 (1973).
33. Henningsson, B., Henningsson, M., Nilsson, T.: Defibration of wood using a white-rot fungus. Res. Note R. 78, Department of Forest Products, Royal College of Forestry, Stockholm, 1972, p. 26.
34. Eriksson, K.-E., Goodell, E. W.: Can. J. Microbiol. **20**, 371 (1974).
35. Eriksson, K.-E.: In Symposium on enzymatic hydrolysis of cellulose. Aulanko, Finland, 12–14 March 1975. SITRA, Helsinki 1975, p. 263.
36. Hartley, R. D., Jones, E. C., King, N. J., Smith, G. A.: J. Sci. Food Agric. **25**, 433 (1974).
37. Kirk, T. K., Moore, W. E.: Wood Fiber **4**, 72 (1972).
38. Moore, W. E., Effland, M. J., Millett, M. A.: J. Agric. Food Chem. **20**, 1173 (1972).
39. Mandels, M., Hontz, L., Nystrom, J.: Biotechnol. Bioeng. **16**, 1471 (1974).
40. Dietrichs, H. H., Hennecke, F. E.: Holz Roh- u. Werkstoff **32**, 13 (1974).
41. Han, Y. W., Callihan, C. D.: Appl. Microbiol. **27**, 159 (1974).
42. Gupta, J. K., Gupta, Y. P., Das, N. B.: Agric. Biol. Chem. **37**, 2657 (1973).
43. Martinez, G. D. V., Ogawa, T., Shinmyo, A., Enatsu, T.: J. Ferment. Technol. **52**, 378 (1974).
44. Peitersen, N.: Biotechnol. Bioeng. **17**, 361 (1975).
45. Peitersen, N.: Biotechnol. Bioeng. **17**, 1291 (1975).
46. Tarkow, H., Feist, W. C.: Advan. Chem. Ser. **95**, 197 (1969).
47. Warwicker, J. O., Jeffries, R., Colbran, R. L., Robinson, R. N.: A review of the literature on the effect of caustic soda and other swelling agents on the fine structure of cotton. Shirley Institute Pamphlet No. 93, 1966, p. 247.
48. Garves, K., Dietrichs, H. H.: Holzforschung **29**, 39 (1975).
49. Mandels, M., Hontz, L., Brandt, D.: Proc. Army Sci., Conf., West Point, New York, 1972. U. S. Nat. Techn. Inform. Serv., Ad. Rep. 750 351 – 00, p. 16.
50. Kato, N., Shibasaki, I.: J. Ferment. Technol. **52**, 177 (1974).
51. Callihan, C. D., Dunlap, C. E.: Construction of a chemical-microbial pilot plant for production of single cell protein from cellulosic wastes. U. S. Environmental Protection Agency, SW-24 c, U.S. Gov. Printing Office, Stock No. 5502 – 0027, 1971, p. 126.
52. Walseth, C. S.: TAPPI **35**, 228 (1952).
53. Katz, M., Reese, E. T.: Appl. Microbiol. **16**, 419 (1968).
54. Dunlap, C. E.: SCP from cellulose. 72nd Nat. Meeting, AIChE, St. Louis, 1972.
55. Ghose, T. K.: Biotechnol. Bioeng. **11**, 239 (1969).
56. Rogers, C. J., Coleman, E., Spino, D. F., Purcell, T. C., Scarpino, P. V.: Envir. Sci. Technol. **6**, 715 (1972).
57. Stone, J. E., Scallan, A. M., Donefer, E., Ahlgren, E.: Advan. Chem. Ser. **95**, 219 (1969).
58. Markkanen, P., Eklund, E.: In Symposium on enzymatic hydrolysis of cellulose. Aulanko, Finland, 12–14 March 1975. SITRA, Helsinki 1975, p. 337.

59. Technical Research Centre of Finland, Biotechnical Laboratory, unpublished results.
60. Humphrey, A. E.: In Symposium on enzymatic hydrolysis of cellulose. Aulanko, Finland, 12–14 March 1975. SITRA, Helsinki 1975, p. 437.
61. Mandels, M., Kostick, J., Parizek, R.: J. Polymer Sci. Part C, **36**, 445, (1971).
62. Li, L. H., Flora, R. M., King, K. W.: Arch. Biochem. Biophys. **111**, 435 (1965).
63. Mitra, G., Wilke, C. R.: Enzymatic utilization of waste cellulosic. Lawrence Berkeley Lab. Rep. 2334, 1975, p. 173.
64. Toyama, N.: Advan. Chem. Ser. **95**, 359 (1969).
65. Wilke, C. R., Yang, R. D.: In Symposium on enzymatic hydrolysis of cellulose. Aulanko, Finland, 12–14 March 1975. SITRA, Helsinki 1975, p. 485.
66. Srivinasan, V. R.: In Symposium on enzymatic hydrolysis of cellulose. Aulanko, Finland, 12–14 March 1975. SITRA, Helsinki 1975, p. 393.
67. Reed, T. B., Lerner, R. M.: Science **182**, 1299 (1974).
68. Linko, M.: Kemia-Kemi **2**, 602 (1975).
69. Nyiri, L. K.: In Symposium on enzymatic hydrolysis of cellulose. Aulanko, Finland, 12–14 March 1975. SITRA, Helsinki 1975, p. 467.
70. Romantschuk, H.: in Single-Cell Protein II. S. R. Tannenbaum, D. I. C. Wang, Eds., MIT Press, Cambridge, Mass. 1975, p. 344.
71. Dunlap, C. E.: Food Technol. 1975, No. 12, 62.
72. Brandt, D.: In Cellulose as a chemical and energy resource. Biotechnol. Bioeng. Symp. No. 5, John Wiley & Sons, New York 1975, p. 275.
73. Wolnak, B.: Present and future technological and commercial status of enzymes. National Science Foundation, Rep. no. NSF/RAX/N–73–002, 1972, p. 44.
74. Nystrom, J. M., Allen, A. L.: In Symposium on enzymatic conversion of cellulosic materials: technology and applications. Biotechnol. Bioeng. Symp. No. 6, John Wiley & Sons, New York 1976, p. 55.
75. Millett, M. A., Baker, A. J., Satter, L. D.: In Symposium on enzymatic conversion of cellulosic materials: technology and applications. Biotechnol. Bioeng. Symp. No. 6, John Wiley & Sons, New York 1976, p. 125.
76. Wilke, C. R., Yang, R. D., Stockar von U.: In Symposium on enzymatic conversion of cellulosic materials: technology and applications. Biotechnol. Bioeng. Symp. No. 6, John Wiley & Sons, New York 1976, p. 155.
77. Seeley, D. B.: in Symposium on enzymatic conversion of cellulosic materials: technology and applications. Biotechnol. Bioeng. Symp. No. 6, John Wiley & Sons, New York 1976, p. 285.
78. Gaden, E. L.: Ch. E. Division Award Lecture, ASEE Nat. Meeting, PPI, June 1974.
79. Miller, D. L.: in Symposium on enzymatic conversion of cellulosic materials: technology and applications. Biotechnol. Bioeng. Symp. No. 6, John Wiley & Sons, New York 1976, p. 307.

Nucleic Acid Damage in Thermal Inactivation of Vegetative Microorganisms

R. F. GOMEZ
Food Microbiology Laboratory, Department of Nutrition and Food Science
Massachusetts Institute of Technology, Cambridge, MA 02139 USA

Contents

1. Introduction

For a number of years the mechanisms of thermal inactivation of vegetative microorganisms have attracted considerable attention. Consequently, the effect of supraoptimal temperatures on various macromolecular components of the cell has been a popular research topic. Some of these include: proteins, DNA, RNA and the membrane. The correlation of the functions that these macromolecules carry out, or better yet, the alteration of these functions by heat, with viability responses has been the most common tool used to study the mechanisms of thermal inactivation. Notwithstanding the many investigations of this nature that have been conducted, a clear picture of the events leading to cellular death has not yet emerged. However, significant progress has been made in elucidating some of the mechanisms by which heat affects macromolecular components and their functions.

In order to convey a clearer understanding of this field it is necessary to define death or inactivation. For the purpose of this review it is defined as the cessation of cell division resulting in the absence of visible colonies on a nutrient plate. This definition leads us into the corollary of injury. Microbial thermal injury, as the term appears in the literature, is used to describe a response, catalyzed by heat, which is manifested by a difference in viable plate counts obtained on two or more different enumeration media. For example, *Staphylococcus aureus* after heat treatment exhibits reduced viable counts on media containing 7% NaCl as compared to a medium without or with less NaCl [1]. Indeed, thermal injury is the most extensively studied thermal response and perhaps it is the approach that has yielded the most significant findings regarding the mechanisms of thermal damage. Implicit in this approach is the assumption that injury is the precursor of thermal inactivation, an assumption with which I take no issue.

The behavior of microorganisms after thermal stress is of great significance to the applied microbiologist. The accurate enumeration of stressed cells in food products, especially pathogens and microorganisms of sanitary significance, is of course of paramount importance. The possibility of not detecting injured pathogens in foods is a problem that concerns food microbiologists. However, research in this area is not only conducted because of the negative aspects of thermal damage on the effectiveness of detection and enumeration procedures of microorganisms, but also because microbial responses to heat can have positive aspects to it. The understanding of the response of microorganisms to heat has the potential of bringing about a more effective use of heat as a processing tool and its use as a synergistic agent with other stresses. Thermal damage offers a powerful system to undertake the study of the fundamentals of cellular damage and repair. In addition, knowledge on the effect of heat on ribonucleic acid (RNA) has provided ways of reducing the nucleic acid content of yeast cells intended for the manufacture of single-cell protein. Clearly then, the study of thermally stressed microorganisms is important from both applied and fundamental standpoints.

It is important to point out that a common microorganism has not been used by all investigators. And even when the same microorganism has been used, the experimental conditions varied from laboratory to laboratory. Cultural age, chemical composition of heating medium and recovery conditions are some of the parameters that can have an impact on the final outcome of the experiment. However, even though this complicates the present task, an attempt to point out common denominators and define basic principles will be made.

Rather than attempt to cover all aspects of thermal damage, I will mostly examine the evidence linking nucleic acid damage to thermal inactivation, and whenever possible, propose and discuss models that account for the available data. In light of the current interest in RNA damage, degradation and resynthesis, as well as deoxyribonuclic acid (DNA) damage and repair in microorganisms, this review is offered.

2. RNA Damage

Although there are three species of RNA (rRNA, mRNA, and tRNA), most, if not all, of the literature dealing with heat stresses is concerned with rRNA. Consequently, this discussion will concentrate on the effects of heat on ribosomes and rRNA. However, it must be recognized that the effects of heat on mRNA and tRNA may be of importance in cell viability.

2.1 Leakage of 260 nm-absorbing Materials

Leakage of cellular components does not necessarily imply that intracellular degradation has occured. Depletion of intracellular pools, without depolymerization, can account for leakage of RNA constituents. In addition, leakage of intracellular components has also been presented as evidence of membrane damage [1, 2]. However, since RNA depolymerization does occur during and after heating (to be reviewed in the next section), leakage of RNA constituents is considered a part of the overall sequence of events after heat treatments and thus pertinent to this discussion.

The earliest available evidence that heat has a marked effect on RNA is the observation that upon heating, cells leak 260 nm-absorbing materials. However, RNA-like materials are not the only cellular components that escape cells upon heating. Other small molecules, such as metals and amino acids, are found in the supernatants of heated cell suspensions [1].

Hancock [3] reported that boiling promoted leakage of 260 nm-absorbing materials from *S. aureus.* Leakage also occurs at lower temperatures, and the leaked material has been identified as RNA constituents [4]. Others [1, 2, 5–8] have confirmed the leakage of 260 nm-absorbing materials from heated suspensions of *S. aureus*.

Leakage of RNA constituents from heated cells has also been reported in a variety of other microorganisms. Hansen and Riemann [9] reported that the loss of stainability of Gram-negative bacteria exposed to heat resulted from a loss of nucleic acids. *Escherichia coli* has been shown to release intracellular 260 nm-absorbing materials upon heating [10, 11]. Similar observations have been made with *Aerobacter aerogenes, Serratia marcescens* [12], *Vibrio marinus* [13], *Streptococcus faecalis* [14], *Pseudomonas fluorescens* [15], and the yeasts *Candida utilis* [10, 16–18], *Candida lipolytica* [19], *Saccharomyces cerevisiae* [20], *Saccharomyces carlsbergensis* [21], and *Candida nivalis* [22].

Although the rates and patterns of leakage vary from microorganism to microorganism, at a given temperature the amount of leakage increases with time and reaches a plateau characteristic of that temperature. The effect of temperature on the degreee of leakage has been studied in various microorganisms. In *S. aureus* Califano [4] reported that leakage was temperature-dependent. Indeed, this would be intuitively expected. However, the nature of the dependency is the point of interest. Russell and Harries [11] reported that in *E. coli* the higher the temperature the greater the rate of leakage in the range of 50° to 60° C, the final amount of leakage always being of the same order of magnitude as that released from boiled suspensions. However, *S. aureus* heated in water leaks more material at 50° C than at 60° C [2, 6–8], although the initial rate of leakage is faster at 60° C than at 50° C [8]. How these observations relate to RNA degradation will be discussed later in this section.

The age and physiological status of cells can have marked effects on the leakage of 260 nm-absorbing materials. Allwood and Russell [8] showed that the amount of leakage of UV-absorbing materials from heated *S. aureus* in water at 50° C and 60° C was the same for both exponential and stationary phase cultures. However, the initial rates of leakage were faster for exponential phase than for stationary phase cultures. In *A. aerogenes*, leakage decreased when cells were starved in phosphate buffer prior to heating [12].

Another parameter that can have an effect on leakage is the composition of the heating medium. There is greater leakage of RNA-like material at 60° C from *S. aureus* cells suspended in 1 M sucrose than form those in water [6, 7]. This observation has been explained in terms of "RNA protection" by coagulated proteins [8]; the amount of coagulated protein being less when cells are heated in sucrose than when heated in water. On the other hand, *E. coli* heated at 50°, 55°, and 60° C in 0.33 M sucrose leaks as much 260 nm-absorbing material as those heated in water [11].

Mg^{2+} reduces leakage of 260 nm-absorbing material in *S. aureus* at 50° C [2]. This could not be attributed to osmotic effects due to the low Mg^{2+} concentration which was used. Instead, the observation was explained by the ability of Mg^{2+} to stabilize ribosomes during heating [23], thus in turn reducing RNA degradation and leakage. Mg^{2+} also prevents excretion of macromolecules from *E. coli* treated with chloramphenicol [24]. This was attributed to a stabilizing effect of Mg^{2+} on the cell envelope.

If leakage is to be taken as an indication of RNA degradation, then it is worthwile to review the nature of the 260 nm-absorbing material that is leaked. This kind of data is scarce. Thermally injured *C. nivalis* leaks 5′-monophosphates without lysing [22]. No polymeric material was found in the heating medium. With *V. marinus* both polymeric and acid soluble RNA-like materials were found in the supernatant of heated cells [13]. Leakage material from heat-shocked *C. utilis* is mainly 3′-mononucleotides [18].

The evidence that leakage is a result of degradation is strong. Good correlative studies have been conducted to show this. Allwood and Russell [8] established a sequence of events in heated *S. aureus* which indicated that RNA degradation was followed by increases in pool size and finally leakage. Similar studies have been conducted with *E. coli* and *C. utilis* [10], *A. aerogenes* [12], and *Ps. fluorescens* [15]. In most of these studies depletion of intracellular pools could not solely account for the amount of leakage observed; thus, degradation had to occur.

How is leakage related to viablility of heated cells? It is best that this discussion be postponed until the evidence of RNA degradation has been presented. Nevertheless, I should like to advance a thought at this point. It is difficult to understand how leakage *per se* could account for loss of viability. The observed leakage appears to be a secondary effect produced by a combination of intracellular depolymerization and loss of the membrane permeability barrier. If the cell is able to restore its permeability functions, it should be able to accumulate the necessary precursors from the medium for macromolecular synthesis and repair thermal damage.

2.2 RNA Degradation

Califano [4] reported that leakage material appearing in supernatants of heated staphylococci was RNA. In a psychrophilic Gram-negative rod, heated in 0.25% NaCl, the leakage material was identified by the orcinol reaction as RNA [5]. Similar experiments have demonstrated the degradation of RNA in heated *Ps. fluorescens* [15]. A study of conditions which accelerated death in thermally stressed *A. aerogenes* indicated that these conditions also accelerated RNA degradation [12]. However, it was suggested that the depletion of RNA was probably not the primary cause of death, but rather death was due to the deleterious effect of a rapid increase in endogenous pool constituents resulting from RNA degradation.

Allwood and Russell [6] indicated that in *S. aureus* leakage was not solely effected by membrane damage, but must be the result of some intracellular changes in the stability of RNA. They suggested that leakage may be the result of RNA breakdown. RNA breakdown products in supernatants of heated *V. marinus* have also been reported [13].

As with leakage of 260 nm-absorbing material, RNA degradation kinetics are dependent on temperature. In *S. aureus*, at 60° C, the initial rate of degradation is faster than at 50° C [8]. However, at 50° C the extent of degradation is larger than at 60° C. This was attributed to a secondary breakdown of RNA due to enzymatic activity on RNA, the enzyme(s) being denatured at 60° C. A comparison of exponentially growing cultures and stationary phase cultures demonstrated that there was little difference in kinetics and extent of RNA degradation between the two types of cultures [8]. Since stationary phase cultures are more heat resistant than exponential ones, the last observation argues against the strict correlation between thermal inactivation and gross RNA degradation. It has also been shown that heating *S. faecalis* causes a reduction of cellular RNA but the percent reduction does not exhibit a 1 : 1 correspondence with percent reduction in viability [14].

In *S. aureus*, sucrose and NaCl in the heating fluid have been shown to accelerate RNA degradation at 60° C and decrease it at 50° C [2, 8]. Mg^{2+} also reduces RNA degradation, presumably by stabilizing ribosomes [2]. However, this reduction in RNA degradation is only observed at 50° C and not at 60° C.

In both *S. aureus* and *S. faecalis*, it is clear that RNA degradation occurs, as evidenced by the rise and fall of RNA in cellular pools [2]. However, degradation extent and kinetics do not always correlate with cellular death.

Various experiments have been designed with *A. aerogenes* to determine the relationship between RNA degradation and cell viability [12]. At 47° C cold acid-soluble UV-absorbing material initially present within the cell increased, indicating degradation, and then decreased due to leakage after a substantial decrease in viability (90%) was observed. Conditions which decreased the death rate, such as the presence of Mg^{2+}, decreased RNA degradation. A variety of other conditions also indicated that RNA degradation was related to the survival characteristics. For example, EDTA accelerated death rate and RNA degradation. It was also observed that death rates and RNA degradation were higher when cells were heated in the presence of KCl. This was attributed to the possible displacement of Mg^{2+} by K^+ and induction of ribosome instability.

One notable exception in the correlative scheme was the observation that heating in distilled water under anaerobic conditions increased the death rate but reduced RNA degradation [12].

If the decreased RNA content in heated cells is responsible for death, then a decrease in RNA prior to heating should lower the thermal resistance. In fact, a starved sulture of *A. aerogenes* which had lost 38% of its RNA were still 95% viable and had a greater heat resistance than non-starved cultures [12]. However, the rate of degradation in starved bacteria was less than in non-starved bacteria. Similar experiments were also conducted with *S. marcescens* and yielded similar results.

The question of RNA content and viability has also been studied during starvation of bacteria. In *A. aerogenes*, RNA was found to be degraded during nutrient starvation [25]. However, the ability to survive starvation does not always correlate with the extent of RNA degradation [26]. In *Sarcina lutea* a 25% loss in RNA was found to occur during starvation without significant losses in viability [27]. Similar results have been reported with *E. coli* [28], *A. aerogenes* [29], *Lactobacillus casei* [30] and *Euglena gracilis* [31]. These results suggested [27] that a decrease in RNA may not be lethal if the mechanism for resynthesis of the polymer from precursors is intact.

Finally, the thermal degradation of RNA has also been reported in yeasts. In *S. cerevisiae,* the ratios of protein to nucleic acids have been kown to increase 10-fold after heat shocking in phosphate buffer [20]. Similar results have been obtained in *C. utilis.* These processes have found practical applications in the reduction of nucleic acids in single-cell protein. Maul *et al.* [16] and Ohta *et al.* [18] have found that cellular RNAase(s) may be activated by a heat treatment of 68° C for a few seconds or by the use of chemicals. When the heat shock process is used, the cells are first heat shocked as they come from the fermentor in spent medium and then incubated at 52.5° C for 2 h. It was established that 1 h at 50° C followed by 1 h at 55° C was the optimal incubation condition after the heat shocking step was performed; however, for a practical situation, the compromise incubation temperature of 52.5° C for 2 h is suitable [16]. The final RNA hydrolysis products from *C. utilis* are 3 -mononucleotides with the RNA content being reduced from approximately 7–8% to 1–2% [18].

2.3 Ribosomal Damage

Ribosomes undergo conformational change upon heating [23, 21–23]. The conformational changes have been demonstrated via:

i) decreases in viscosity,
ii) hyperchromicity,

iii) optical rotation, and
iv) buoyant density studies.

The conversion of ribosomes to heated particles via an intermediate stage of open structures has been shown [23]. The open structures consist of rRNA to which ribosomal proteins are attached. Thermal treatment of ribosomes, in the range of 40° to 70° C, resulted in a decrease of sedimentation velocity [23]. Examination of the heated particles revealed that the ribosomal RNA remained intact and that they retained almost all of their proteins. In addition, it was suggested that the integrity of the rRNA determines the sedimentation pattern of the ribosomes after exposure to elevated temperatures and that the secondary structure of the ribosomal protein varies less with temperature than the secondary structure of the rRNA moiety [23]. These two suggestions lead to the conclusion that the dominant factor in thermally induced ribosomal changes is the rRNA. However, this is not to be taken as a suggestion that ribosomal proteins do not play a role in ribosomal integrity. Indeed, it has been put forward that the stabilization of double-helical secondary structure within ribosomal particles arises from the interaction with ribosomal proteins [35].

It may be worthwhile to note that subribosomal particles of *Bacillus stearothermophilus*, a thermophile, are more stable to heating, by approximately 10° C, than those of *E. coli* [36, 37]. A correlation between ribosome stability and maximum growth temperature has also been established [38, 39]. In addition, the functionality of ribosomes in protein synthesis is more heat resistant in *B. stearothermophilus* than in *E. coli* [34]. However, ribosome functionality of *E. coli* is observed at temperatures well above the optimum for growth [34]. The amount of polyphenylalanine synthesized by *E. coli* ribosomes was constant from 37° C to 55° C. At 60° C the activity of the ribosomes was halved and, furthermore, it was found that *E. coli* ribosomes could withstand a 10% loss in RNA double-helical secondary structure without losing activity. *A priori* this would tend to discredit the relationship between ribosome damage and thermal inactivation. However, it must be noted that these experiments were conducted with purified ribosomes free of RNAases. It has been shown that RNAases present in ribosomal preparations will result in degradation upon heating [35, 36, 39–46]. Of course, RNAases are present *in vivo* and the action of these on heated ribosomes may be the event responsible for RNA degradation at high temperatures rather than heat-catalyzed RNA hydrolysis. The characteristics of bacterial and yeast RNAases along with possible mechanisms of degradation will be discussed in the next section. For now, further evidence of ribosomal damage in heated microorganisms will be provided. Using various assay procedures of thermal injury it has been demonstrated that the expression of injury is accompanied by specific ribosomal damage. In *S. aureus* the sensitivity to NaCl [47] after heat treatments is congruent with ribosome damage [48]. Similar studies have shown that sensitivity of *S. typhimurium* to Levine Eosin Methylene Blue Agar containing 2% NaCl [49] and of *Ps. fluorescens* to trypticase soy agar [15] coincide with ribosomal damage. In heated *S. aureus* [48], methylated albumin kieselguhr column analyses demonstrated that at least one of the sites of thermal injury was the ribosomal ribonucleic acid. Sucrose gradient analyses confirmed the loss of 70S ribosomal particles. The 30S subunit appears to be selectively degraded [50]. Polyacrylamide gel electrophoresis demonstrated that the 16S RNA was lost and the secondary structure of the 23S was altered. The latter alteration was shown by the susceptibility of heated 23S RNA to pancreatic ribonuclease [50].

Heated Cells of *S. typhimurium* completely lost their 16S RNA species but the 23S RNA was only partially lost [49]. In *Ps. fluorescens* [15], both 23S and 16S species were affected, but only partially lost.
In other systems where ribosome degradation has been studied, such as during starvation of *Ps. aeruginosa* in saline [51], it has been shown that the 50S particle is preferentially degraded. On the other hand, in *E. coli* [52] both the 23S and the 16S RNA are degraded during nutrient starvation and heat treatment at 50° C. These differences may point to the different RNA degrading mechanisms that are triggered after heat treatments. Furthermore, there are indications that the suspending medium may have a marked effect on the pattern of degradation. Haight and Ordal [53] suggested that buffer systems may dictate which degrading enzyme predominates.

2.4 RNA Degrading Enzymes

There are three major groups of enzymes known to depolymerize RNA: phosphotransferases, phosphodiesterases, and phosphorylases. Their specificities, characteristics and the possible biological roles of these enzymes have been reviewed elsewhere [54]. The discussion of these enzymes will be restricted to their role in thermal injury and inactivation. An attractive hypothesis is to presume that heat creates conditions in which the catabolism of RNA is favored over synthesis. The question is, how does this happen?
First, the conditions which are necessary for thermally induced degradation of RNA need to be established. The enzyme must be active, that is, not denatured by heat, it must have its ion requirements satisfied and the substrate must be available to it.
The thermal stability of ribonucleases is a widely accepted fact. For example, pancreatic ribonuclease has been shown to be most active at 55° C against Candida yeasts [17]. Thermal stability of these enzymes is best evidenced by the observed enzymatic degradation of RNA in heated bacteria and yeasts, and the degradation of *S. aureus* ribosomes at elevated temperatures [53].
Cation requirements vary for the different types of enzymes. The cyclizing RNAases have no requirement for divalent cations [54]. On the other hand, the phosphodiesterases active on RNA require either Mg^{2+} or Ca^{2+}. The liberation and activation of ribosomal RNAase by treatment with 0.3 M sodium chloride has been reported [55]. Furthermore, the addition of Na_2HPO_4 to the heating suspension greatly accelerates the reduction of nucleic acids in *S. cerevisiae* [20]. Similar results have been obtained with NaCl [56]. These effects of sodium may provide an explanation for the sensitivity of heated bacteria to salts [47]. It should be remembered that the influence of ions on polynucleotide-degrading reactions is rather complex. Changes in the ion concentration can considerably modify the structure of the RNA molecule [57–60]. This can markedly affect the accessibility of substrate to the enzyme or the release of products from cleaved RNA molecules. These effects are quite distinct from interactions of ion with the enzyme-substrate complex.
It has been reported that the structure of ribosomal RNA changes with heat and only the portion of the RNA which suffers structural changes can be hydrolyzed by activated RNAase [35]. In intact polysomes rRNA is not affected by levels of RNAase which readily attack ribosomal subunits or the interribosomal segments of mRNA [61–63].

Ribosomes may even bind RNAase without hydrolysis of their RNA [35, 55, 64–66]. This resistance is presumably due to both complexing of rRNA and steric inaccessibility.
A very important effect of heat could be the activation of the enzymes themselves. This could occur via the denaturation of a heat-labile inhibitor. This is illustrated in the case of alkaline RNAase in mammalian tissue cells [67–69]. The inhibitor is a protein which has been purified and characterized. Protein inhibitors of RNAases have also been observed in bacteria [70] and in yeasts [19, 71]. Provided that ribosomes contain a latent ribonuclease the following hypothesis seems to be reasonable. Elevated temperature may initiate RNA degradation by two simultaneous events: 1) heat changes the conformation of the ribosome and makes them more susceptible to the attack of RNAase; 2) the conformational change leads to a release of the ribosome bound RNAase and/or heat degradation of a labile inhibitor, thus rendering the enzyme active.
In *C. utilis* heat shock apparently does not function by denaturing ribosomes, since native ribosomes in homogenates from untreated cells can serve as substrates for the heat-activated ribonuclease [18]. However, it is possible that the homogenization procedures may have affected the ribosomes to make them susceptible to enzymatic degradation.
I would like to propose that the technique of thermal injury and its effect on RNA may prove useful in studies of mechanisms of RNA regulation. It has been suggested that in the case of rRNA, it is not the synthesis of rRNA that is controlled, but rather that the level of rRNA in a cell is a function of protective proteins [72]. These protective proteins could be of a ribosomal nature, protecting RNA by steric hindrances, or of a more specialized nature, such as inhibitors. Thermal shocks, as already mentioned, could distrub the balance of active *versus* inactive RNAases and catalyze RNA breakdown. The usefulness of thermal shocks as a tool in the study of rRNA regulation would come from its ability to disturb the equilibrium between synthesis of RNA and degradation.

2.5 RNA Resynthesis

Can rRNA be resynthesized, and how does resynthesis correlate with viability? After thermal insults, *S. aureus* is unable to reproduce on media containing 7.5% NaCl [1]. When the heated bacteria are placed in a complete recovery medium they regain their salt tolerance. This has been the basis for studying the events which are necessary for recovery.
In *S. aureus* it has been reported that chloramphenicol, penicillin and cycloserine do not affect normal recovery (regaining of salt tolerance) but actinomycin D brings about inhibition [1]. These results suggested that while protein and cell wall synthesis are not required for repair, RNA synthesis is. The requirement of RNA synthesis was further supported by the incorporation of uniformly labelled ^{14}C-aspartic acid and glycine into nucleic acid and soluble pools of recovering cells [1]. Furthermore, the addition of actinomycin D to the recovery medium retards the return of normal metabolic activity [73]. However, in the latter case it was also observed that chloramphenicol retarded the reconstitution of metabolic activity. A more recent report [76] indicated that full recovery of salt tolerance by *S. aureus* in the presence of chloramphenicol does not occur. Thus, the requirement for protein synthesis during repair of *S. aureus* needs further investigation.

In previous sections it was indicated that heat treatments caused a loss of rRNA in *S. aureus* [48] and that this was also most pronounced in the 16S RNA species [50]. During recovery of the heated bacteria the rRNA is resynthesized [48]. Even though the 30S species is specifically degraded during heat treatments, both ribosomal particles (30S and 50S) are reassembled [74]. This is indicative of turnover of the 50S particles and of conservation and recycling of the ribosomal proteins.

With *S. faecalis* the reacquisition of tolerance to 6% NaCl has been shown to be dependent on RNA synthesis [75]. In this case only 0.26% of the bacteria recovered their salt tolerance in the presence of actinomycin D. It was also concluded that protein synthesis was not involved in repair. However, careful examination of the data indicates that chloramphenicol inhibited repair of approximately 60% of the cells. It was argued that chloramphenicol was "slightly" lethal to control bacteria, enumerated on media containing salt, and that this was probably the reason why complete recovery was not observed in heat-injured cells. Although the conclusion that protein synthesis is not required for repair of *S. faecalis* holds true for 40% of the bacteria, the argument for the complete independence of repair from protein synthesis is not convincing. Nevertheless, our interest lies with the requirement of RNA synthesis during repair of *S. faecalis,* and this is indeed an inescapable fact.

In *Ps. fluorescens* [15], rifamycin inhibited repair of injured cells. In addition, resynthesis of RNA corresponded with repair of thermal injury and inhibition of RNA synthesis with rifamycin-inhibited repair. Furthermore, both mature 23 and 16S RNA species were resynthesized during repair. An accumulation of 17S RNA prior to the appearance of 16S RNA indicated the presence of a temperature-sensitive step in the maturation process of 16S RNA. A puzzling event was observed immediately after thermal stress. During the first hour of recovery, injury as assayed by plate counts continued to occur, and yet both 23 and 16S RNA were being resynthesized. The authors argued that this may be due to the dominance of another type of lesion during the first hour after thermal treatment independent of RNA degradation. I would like to add to that explanation. In the first hour of repair, degradation was still occuring, as shown by the leakage of 260 nm-absorbing material. It is possible that a threshold value of RNA content to ascertain tolerance to the enumeration media had not yet been reached. Only when resynthesis overtakes degradation will the bacteria show repair of thermal injury.

Thermal injury to *S. typhimurium* and repair has been shown to correspond to RNA degradation and resynthesis, respectively [49]. In addition to RNA synthesis, it was shown that the repair process was dependent on adenosine triphosphate and protein synthesis. The conclusions were arrived at on the weight of evidence indicating that 2,4-dinitrophenol and chloramphenicol inhibited repair. RNA synthesis was inhibited by rifamycin and 5-fluorouracil. With this microorganism, it was suggested [77] that maturation of the newly synthesized rRNA was the rate limiting step rather than total RNA synthesis during recovery. In addition, very little old ribosomal protein was found associated with newly resynthesized rRNA. Instead, the new ribosomes contained newly synthesized proteins as well as new rRNA.

Finally, repair of thermal injury in *Bacillus subtilis* also corresponds with RNA synthesis and is blocked by actinomycin D [78]. Both rRNA species (16S and 23S) are resynthesized during repair. Unlike the inhibitory effect of chloramphenicol on the maturation

process in *S. typhimurium* [77], in *B. subtilis* maturation was not affected by chloramphenicol.
One negative statement on the involvement of RNA synthesis during repair has been made [79]. It was claimed that the involvement of RNA in repair of thermally damaged *E. coli* could not be direct since chloramphenicol inhibited repair without inhibiting RNA synthesis. This conclusion was unwarranted. The data indicated that indeed protein synthesis was required, but it told nothing about RNA synthesis. Both events, RNA and protein synthesis, could be required in *E. coli*.

2.6 RNA Damage vs. Cell Viability

Microorganisms upon heat treatment leak 260 nm-absorbing material and this material is derived from polymeric RNA. In addition, depending on the microorganisms, ribosomal subunits are specifically degraded and the RNA component and not the protein component is the one attacked. These changes in RNA content and ribosomal units have been correlated with biological inactivation and thermal injury, where thermal injury is assayed as a sensitivity to an enumeration medium. The enumeration media in which injured microorganisms are not able to reproduce commonly contain 3 to 7% NaCl, or some other type of selective agent.
During the repair process, in which microorganisms regain their tolerance to the selective agent, RNA synthesis is observed along with the regeneration of the ribosomal subunits lost during heating. Here again a correlation between repair of thermal injury and RNA synthesis is observed. Whether RNA degradation and synthesis are causes or consequences of thermal injury and repair, respectively, is a much discussed subject.
There is little doubt that a bacterium without a full complement of RNA, necessary to meet environmental changes, cannot survive, or rather, cannot initiate growth and division until the RNA has been regenerated. This thought of course is of little help in determining whether RNA synthesis is the primary site of damage. The term "primary" suggests that there may be other sites, and indeed this seems to be the case, e.g. membrane damage; in addition it suggests that RNA damage is the dominant damage. This latter suggestion, without belittling the role of RNA damage, is dangerous. Let us take a hypothetical situation in which 3 sites are damaged by heat and assume that all three sites or functions performed by these sites are essential for growth. In order to obtain fully viable, uninjured cells, it would be necessary to replenish or repair all three sites. Repair of one or two would not be enough. I put forth this argument since heat is a multitarget agent [80] and affected sites include not only RNA but also membrane, proteins and DNA.
It is my opinion that RNA damage is indeed an important event during heat treatment and that it may very well be the limiting factor in repair, but that RNA damage can solely account for viability behavior of heat injured cells I do not accept.
It is interesting to consider the effect of NaCl on RNA resynthesis. It cannot be unequivocally established whether NaCl inhibits RNA resynthesis in thermally injured cells, or if it affects other functions that indirectly prevent resynthesis. NaCl has two effects on heated *S. aureus*. One is seen immediately after heat treatment when it proves to be

bactericidal, and the other after 30 min of incubation in non-salt medium when it is only inhibitory [47]. This observation could reflect the involvement of the permeability barrier in determining the level of NaCl allowed into the cells. A rapid repair of the membrane could limit the concentration of sodium ions to an inhibitory level.
I should like to propose that further experimentation be conducted in the tradition of Tomlins and Ordal [49, 77], where subtle changes in ribosomal components are followed, rather than gross effects such as degradation and leakage. The judicious use of temperature sensitive mutants defective in various steps of RNA synthesis could be fruitful. Again I reiterate that these procedures would yield valuable information concerning the effect of heat on rRNA, and may in fact provide a useful tool in understanding the mechanisms of rRNA regulation.

3. DNA Damage

The effects that ionizing and non-ionizing radiation have on the DNA molecule have been the subject of extensive experimentation. However, other environmental stresses such as temperature have been virtually overlooked. In this section, the evidence that points to the DNA molecule as an important target of heat treatment will be presented. As early as 1956, Wood [81] alluded to the possibility that at least a portion of thermal inactivation acts through damage to the genetic apparatus. He arrived at this conclusion by observing that diploid forms of yeasts were more heat-resistant than haploid forms, and suggested that heat and X-irradiation acted at similar sites.

3.1 Heat as a Mutagen

If mild heat (48° C to 60° C) has any effect on the DNA molecule, it could be expected that stable mutations could result from thermal exposure. Zamenhof and Greer [82] demonstrated that heat treatments in 50% sucrose solutions (60° C for 2–4 h) substantially increased mutation frequencies from prototrophy to stable auxotrophy in *E. coli* over control cultures. Subsequently, it was shown [83] that heating dry *B. subtilis* spores also resulted in stable auxotrophic mutations. It was believed [83] that the functions of sucrose or desiccation was to increase the heat resistance of cells to a point where enough heat could be supplied in order to cause changes in the DNA molecule without significantly affecting viability.
Recently it has been shown [84] that exposure to 52° C induces mutations in both nuclear and cytoplasmic DNA in a haploid strain of *S. cerevisiae.* These authors showed that immediately after heat treatment, the fraction of "petite" mutants in the population increased from less than 0.01 to about 0.60 and that of canavanine-resistant mutants from less than 10^{-5} to about 8 x 10^{-4}. Mutation frequencies were maximal for cells in the exponential phase exposed for 20 min at 52° C. Later, it was shown [85] that if, after heating and before application of selective conditions, yeast cultures were held in water at 20° C the mutation frequencies decreased, and their survival was greatly enhanced. This decrease in mutation frequency and increase in survival was interpreted

as repair of heat injury and was found to be blocked if incubation was carried out at 4° C or in the presence of protein synthesis inhibitors.
Even though the available information on heat induced mutations is limited, it would appear that heat can be a mutagenic agent and therefore has an effect on the DNA molecule. However, it must be pointed out that this effect does not have to be a direct one. For example, mutations could be induced by either the activation, inactivation or modification of enzymes involved in the synthesis and maintenance of the DNA molecule.

3.2 Genetic Evidence

Various workers have attempted to compare the radioresistance of organisms which are radiation-sensitive, and thus defective in DNA repair systems, with their heat-resistance. Bridges *et al.* [86, 87], found a general correlation between the sensitivity of *E. coli* cells to gamma radiation and heat stress. Pauling and Beck [88] have shown that a DNA-ligase-defective mutant of *E. coli* is markedly more sensitive to heating than the wild type. However, Schenberg-Frascino and Moustacchi [84] examined the thermo-sensitivity of six different radiosensitive cultures of *S. cerevisiae* and found that only one culture showed a greater sensitivity to heat than the corresponding wild type. In contrast, Matsumoto and Kagami-Ishi [89] found a general correlation with both *E. coli* and *S. cerevisiae.* Finally, a *S. typhimurium* carrying an R-factor conferring increased resistances to radiation and alkylating agents was shown to be more heat sensitive than the wild type parent [90]. Thus, it was concluded that increased repair capacities of bacteria containing R-Utrecht plasmids do not extend to include repair of thermal damage.
Although the positive evidence points to a possible relationship between repair systems for thermal and radiation lesions, it is obvious and expected that both agents are not identical. The correlation is further complicated by the fact that, depending on the time-temperature relationship, other lethal events can take place during thermal treatment, e.g. RNA degradation, which could mask or become dominant over DNA damage. For example, Haynes [91] found that at 60° C strains B/r, B and Bs-1 of *E. coli* were equally sensitive to heat. In contrast, Bridges *et al.* [86, 87] found that B/r was more resistant to 52° C than Bs-1.
At any rate, since some mutants deficient in DNA repair systems are more heat-sensitive than the wild types parents, it seems reasonable to assume that survival of some heat-treated microorganisms largely depends on their ability to repair their DNA or that the missing functions have additional roles other than DNA repair and that these roles are essential for repair of thermal damage.

3.3 Biochemical Evidence

One means of assessing damage to the DNA molecules is the sucrose alkaline gradient technique of McGrath and Williams [92]. This technique measures indirectly the number of scissions of the DNA which may be caused by hydrolysis of the phosphodiester link [93, 94].
Bridges *et al.* [87] showed that DNA from *E. coli* B$_3$r and Bs-1. heated at 52° C for 20

and 40 min sedimented through alkaline sucrose gradients at faster rates than DNA from unheated bacteria. However, these authors could not demonstrate any repair of DNA breaks after heat treatment as it occurs after X-ray damage [92]. Dugle and Dugle [95] demonstrated that heat treatment of *B. subtilis* at 75° C resulted in both single and double-stranded DNA breakage. Woodcock and Grigg [96] also reported that exponentially growing cultures of *E. coli* B exhibited single and double-stranded DNA breaks when exposed to 52° C, and furthermore, that the breaks were repaired during subsequent incubation in phosphate buffer at 37° C. In addition, during the repair period, viability increases were observed. These increases in plate counts were interpreted to be due to repair of thermal injury and not new growth.

It is important to note that heat treatments do not always result in DNA breakage. Sedgwick and Bridges [97] reported that strains *E. coli* H/R 30-R *res*$^+$ and R15 *res*A did not show any significant strand breakage at 52° C. However, the *res*A mutation in the *E. coli* family (similar to *pol*A in *E. coli* K-12 having this same map position) does impart thermal sensitivity [86]. On the other hand, they did show an increase of acid-soluble radioactivity. They interpreted this to mean that DNA had undergone extensive exonucleolytic hydrolysis during heating instead of single-strand breaks.

Since *res*A cells are more heat sensitive than *res*$^+$ cells, but neither shows any DNA breakage [97], it could be concluded that the repair function performed by the *res* gene product on thermal lesions, and therefore its conferring of resistance, does not involve the rejoining of DNA strands, or that exonucleolytic attack in these strains precludes the detection of strand breakage.

Sedgwick and Bridges [97] also reported that *pol*$^+$ and *pol*A [98] show DNA breakage after heating, although both strains behaved similarly regarding the extent of DNA breakage and survival. The authors concluded that the gene product of *pol*A is not responsible for repair of thermal lesions or that it is inactivated at 52° C.

Uptake of radioactive thymine and thymidine has been studied in order to determine the extent of DNA replicative repair after thermal treatments. Post-heat treatment incorporation of labelled precursors has not been reported during the repair periods of *E. coli* and *S. typhimurium* [77, 96]. However, these results do not conclusively demonstrate the absence of DNA repair synthesis. For example, the transport systems necessary for uptake of thymidine into polymeric DNA may be heat-sensitive and the cell may be using endogenous precursors derived from nucleolytic action. Alternatively, the repair of heat lesions may just involve a ligating activity or simply a renaturing of the DNA duplex. However, there is evidence [99] that hydroxyurea blocks the repair of thermal injury. Hydroxyurea is a DNA synthesis inhibitor which acts by inactivating the ribonucleotide diphosphate reductase enzyme system [100, 101]. Thus, is would seem that some DNA synthesis is, after all, required. This conclusion of course assumes that the effect of hydroxyurea is solely that reported by Krakoff *et al.* [100] and Sinha and Snustad [101]. Conversely, nalidixic acid does not inhibit repair of thermal injury in *S. typhimurium* [102]. Nalidixic acid is believed to inhibit DNA synthesis at or near the replicating point [103]. It has also been suggested that it acts by interfering with normal DNA-membrane association [104]. It should be noted that the use of selective inhibitors *in vivo*, without a full understanding of their action, is only a preliminary step in the process of elucidating the mechanisms of DNA damage and its repair.

3.4 DNA Damage and Viability: Mechanisms

The correlation between viability and DNA strand-breakage has to take into account that in viability analyses the cells are exposed to a growth medium after heat treatment, while in sedimentation analyses they are not. In addition, repair processes most likely occur during suitable incubation after heat treatment and, furthermore, DNA damage, as measured by strand breakage may not be expressed until after heat treatment and subsequent incubation. For example, the phenomenon of "minimal medium recovery" (loss of colony forming ability on complex media) of heated *S. typhimurium* [99] is due to exposure to high temperatures followed by incubation in complex media [105]. Minimal medium recovery (MMR) has also been shown [106] to occur in irradiated excision deficient mutants of *E. coli.* DNA breaks in heated *S. typhimurium* (50° C for 15 min) appear after exposure to complex media but not before [105]. It is important to note that repair of thermal injury as measured by the return of the ability of bacteria to grow on complex media coincided with the gradual disappearance of complex media-induced DNA breakage. If conditions are provided in which DNA breaks can be repaired, viability is not lost. Such a condition is provided when heated cells are exposed to complex media under anoxic conditions [107]. In this case, exposure to complex media still results in the initial appearance of DNA breaks; however, they are completely repaired in 150 min. It is noteworthy that the polymerase-deficient *E. coli* strain *pol*A_1 is sensitive to nutritionally complex media of it is previously grown on a minimal medium without the action of heat [108]. I have also found (unpublished results) that *rec* and *uvr*B mutants of *S. typhimurium* are sensitive to a shift from minimal medium to complex media. This could reflect enzymatic requirements, involved in DNA metabolism, for survival following a nutritional imbalance. Along the same lines, there is evidence that complex media will affect mutation frequencies. In one case [109] it has been found that analyses of mutagenized cultures of *S. cerevisiae* on complex media yields higher mutation frequencies than on minimal medium. This is due to a decreased number of survivors on complex media, since mutation frequencies are expressed with reference to the number of survivors. In the second case [110], it has been shown that nutrient broth increases the number of mutants obtained after UV treatment of *E. coli.*

In the interest of completeness, it should be noted that the effect of complex media could be a more specific one than induction of nutritional imbalance. Traces of chemical contaminants in these media could play inducing or inhibitory roles in DNA damage and repair, respectively. Such effects have been reported for a number of different agars [111] and media [112] which seem to inhibit repair of ultraviolet induced lesions in *E. coli.*

It is the consensus of the workers in this field that ultimate DNA strand breakage after mild heat treatment is caused by enzymatic action. Therefore, the following discussion will use this basic premise.

Harris [113] has proposed that enzymes which are used in recombination and repair of DNA have a definite role during normal growth. These enzymatic activities are endonucleolytic, exonucleolytic, DNA polymerizing and ligating. I propose that the activity or control of enzymes carrying out nucleolytic action and/or repair is altered during heat treatment and possibly during postheat treatment incubation, causing an imbalance in rates of degradation and resynthesis.

How could the imbalance occur? Either the substrate (DNA) or the enzymes themselves are altered. Let us examine how the substrate can be changed. Inman [114] has reported that at temperatures of 50° to 55° C, λphage DNA develops denatured regions in 10 min. This is in line with the time-temperature relationships observed in studies on heat damage. It is likely that the regions where denatured sites appear have the richest sequences of the bases adenine and thymine. It should be noted that locally denatured tracks are probably found throughout the DNA molecule during normal growth and it can be supposed that heat treatment enlarges and aggravates these "temporary regions". It has been proposed that "bubbles" or denatured regions occur during normal DNA replication [115]. A model for superhelical DNA in which there are regions of altered secondary structure containing unpaired bases has been proposed [116]. It has been hypothesized that during transcription DNA unwinds [113, 117, 118].

It has been suggested by Woodcock and Grigg [96] that denatured regions could be recognized by endonucleases and breaks introduced. Enzymes that can attack these regions have been found in *E. coli* [119]. Endonucleases which recognize depurinated regions of DNA have also been demonstrated [120].

It is also possible that heat could affect the activity of the enzymes themselves. For example, a situation similar to that occuring in *C. utilis* could arise where an inhibitor of ribonuclease is degraded by heat treatment, allowing ribonuclease to act on RNA [19].

Whether activation or induction of nucleases, if any, is achieved by heat alone or by heat coupled with other factors may depend on the physiological and genetic make-up of the cells. With *S. typhimurium* LT-2 it is necessary to expose heated bacteria to nutritionally complex media in order to create DNA breakage [105]. Therefore, something other than heat is required to cause strand breakage. In contrast, in *E. coli* B [96] breaks appear during heat treatment. Furthermore, it should be understood that activation or induction of nucleases, if it occurs, does not exclude the possibility of DNA structure alteration and *vice-versa.* It is possible to visualize a sequence of events leading to DNA breakage which includes all of the alternatives mentioned.

Regardless of the type of damage exerted by heat, it is clear that repair takes place. Then how does repair proceed in terms of the proposed alternatives? If partial denaturation occurs, then repair means renaturation. The question remains as to what is required for repair of thermal lesions *in vivo* if they are expressed in terms of local denaturation in DNA of bacteria. So far this question cannot be answered. Whether a physical rearrangement of the strands or a more extensive repair system is required is not known. The fact that hydroxyurea blocks the repair of heat injury is indicative of the latter hypothesis. If heat injury means a degradation of inhibitors of nucleases, then resynthesis of these inhibitors is required. There is evidence available [102] that the inhibition of protein synthesis with chloramphenicol prevents the repair of heat injury in *S. typhimurium,* using the MMR phenomenon as an assay system. In addition, exposure of heated *S. typhimurium* to the RNA synthesis inhibitor rifampin causes loss of viability of heated cells and DNA breakage. Repair could also take the form of strand rejoining. Finally, it should be pointed out that the repair systems themselves could be heat inactivated and therefore would have to be rebuilt.

It would be of little substantive value of speculate further as to the repair mechanisms.

Elucidation of these mechanisms must await a clearer understanding of the type(s) of damage exerted by heat on the DNA molecule and its maintenance systems.

The action that physical stresses, such as heat, have on bacterial cells is undoubtedly multifold. However, there is now evidence indicating the genetic appartus to be one of the systems affected, and in many cases its damage can be held responsible for injury behavior and irreversible inactivation of microorganisms.

It has been demonstrated that heat can cause stable phenotypic changes through mutational events and the frequency of these mutations is a function of the degree of damage inflicted on the cell. The observations with mutants defective in DNA repair after heat treatment, and the fact that DNA strand breakage occurs in heated cells, can account for at least some of the injury and death of heated microorganisms.

The basic model for DNA damage caused by heat stresses is based on the premise that the balance between synthetic and degradative systems is altered in stressed cells. This imbalance, I propose, can be caused by changes either in the DNA molecule *per se* or the control of nucleolytic and/or repair enzymes.

Finally, it is clear that the extent of the involvement of DNA in the reversible and irreversible inactivation of bacteria depends not only on the severity of the stress, but also in the genetic and/or physiological make-up of the cell.

Acknowledgements

This investigation was supported by grant FD-00530 from the Food and Drug Administration and is contribution No. 3008 from the Department of Nutrition and Food Science, Massachusetts Institute of Technology, Cambridge, Ma. 02139, U. S. A.

References

1. Iandolo, J. J., Ordal, Z. J.: J. Bacteriol. **91**, 134 (1966).
2. Allwood, M. C., Russell, A. D.: J. Pharm. Sci. **59**, 180 (1970).
3. Hancock, R.: Biochim. Biophys. Acta **28**, 402 (1958).
4. Califano, L.: Bull. World Health Org. **6**, 19 (1952).
5. Moats, W. A.: J. Dairy Sci. **44**, 1431 (1961).
6. Allwood, M. C., Russell, A. D.: Appl. Microbiol. **15**, 1266 (1967a).
7. Allwood, M. C., Russell, A. D.: Experientia **23**, 878 (1967b).
8. Allwood, M. C., Russell, A. D.: J. Bacteriol. **95**, 345 (1968).
9. Hansen, N. H., Riemann, H.: J. Appl. Bacteriol. **26**, 314 (1963).
10. Shibasaki, I., Tsuchido, T.: Acta Alimentaria **2**, 327 (1973).
11. Russell, A. D., Harries, D.: Appl. Microbiol. **15**, 407 (1967).
12. Strange, R. E., Shon, M.: J. Gen. Microbiol. **34**, 99 (1964).
13. Haight, R. D., Morita, R. Y.: J. Bacteriol. **92**, 1388 (1966).
14. Beuchat, L. R., Lechowich, R. V.: Appl. Microbiol. **16**, 772 (1968).
15. Gray, R. J. H., Witter, L. D., Ordal, Z. J.: Appl. Microbiol. **26**, 78 (1973).
16. Maul, S. B., Sinskey, A. J., Tannenbaum, S. R.: Nature **228**, 181 (1970).
17. Castro, A. C., Sinskey, A. J., Tannenbaum, S. R.: Appl. Microbiol. **22**, 422 (1971).

18. Ohta, S., Maul, S., Sinskey, A. J., Tannenbaum, S. R.: Appl. Microbiol. **22**, 415 (1971).
19. Imada, A., Sinskey, A. J., Tannenbaum, S. R.: Biochim. Biophys. Acta **268**, 674 (1972).
20. Canepa, A., Pieber, M., Romero, C., Toha, J. C.: Biotechnol. Bioeng. **14**, 173 (1972).
21. Lee, T. C., Lewis, M. J.: J. Food Sci. **33**, 124 (1968).
22. Nash, C. H., Sinclair, N. A.: J. Microbiol. **14**, 691 (1968).
23. Tal, M.: Biochem. **8**, 424 (1969).
24. Gupta, R. S.: Antimicr. Ag. Chemother. **7**, 748 (1975).
25. Postgate, J. R., Hunter, J. R.: J. Appl. Bacteriol. **26**, 295 (1963).
26. Harrison, A. P., Lawrence, F. R.: J. Bacteriol. **85**, 742 (1963).
27. Burleigh, I. G., Dawes, E. A.: Biochem. J. **102**, 236 (1967).
28. Dawes, E. A., Ribbons, D. W.: Ann. Rev. Microbiol. **16**, 241 (1962).
29. Strange, R. E., Dark, F. A., Ness, A. G.: J. Gen. Microbiol. **25**, 61 (1961).
30. Holden, J. T.: Biochim. Biophys. Acta **29**, 667 (1958).
31. Blum, J. J., Buetow, D. E.: Exp. Cell Res. **29**, 407 (1963).
32. Doty, P., Boedtker, H., Fresco, J. R., Haselkorn, R., Litt, M.: Proc. Nat. Acad. Sci. USA **45**, 482 (1959).
33. Cotter, R. T., McPhie, P., Gratzier, W. B.: Nature **216**, 864 (1967).
34. Cox, R. A., Pratt, H., Huvos, P., Higginson, B., Hirst, W.: Biochem. J. **134**, 775 (1973).
35. Ohtaka, Y., Uchida, K.: Biochim. Biophys. Acta **76**, 94 (1963).
36. Stenish, J., Yang, C.: J. Bacteriol. **93**, 930 (1957).
37. Friedman, S. M., Axel, R., Weinstein, I. B.: J. Bacteriol. **93**, 1521 (1967).
38. Saunders, G. F., Campbell, L. L.: J. Bacteriol. **91**, 332 (1965).
39. Pace, B., Campbell, L. L.: Proc. Nat. Acad. Sci. USA **57**, 1110 (1967).
40. Hall, B. D., Doty, D.: J. Mol. Biol. **16**, 473 (1959).
41. Zubay, G., Wilkins, M. H. F.: J. Mol. Biol. **2**, 105 (1960).
42. Schlessinger, D.: J. Mol. Biol. **2**, 92 (1960).
43. Tamaoki, T., Miyazawa, F.: J. Mol. Biol. **17**, 537 (1966).
44. Wolfe, F. H., Kay, C. M.: Biochem. **6**, 2853 (1967).
45. Leon, S. A., Brock, T. D.: J. Mol. Biol. **24**, 391 (1967).
46. Wolfe, A. D.: Fed. Proc. **27**, 3329 (1968).
47. Erwin, D. G., Haight, R. D.: J. Bacteriol. **116**, 337 (1973).
48. Sogin, S. J., Ordal, J.: J. Bacteriol. **94**, 1082 (1967).
49. Tomlins, R. I., Ordal, Z. J.: J. Bacteriol. **105**, 512 (1971).
50. Rosenthal, L. J., Iandolo, J. J.: J. Bacteriol. **103**, 833 (1970).
51. Gronlund, A. F., Campbell, J. J. R.: J. Bacteriol. **90**, 1 (1965).
52. Kaplan, R. Apirion, D.: J. Biol. Chem. **250**, 3174 (1975).
53. Haight, R. D., Ordal, Z. J.: Can. J. Microbiol. **15**, 15 (1969).
54. Barnard, E. A.: Ann. Rev. Biochem. **38**, 677 (1967).
55. Schell, P. L.: Nature **210**, 1157 (1966).
56. Lindblom, M., Mogren, H.: Biotechnol. Bioeng. **16**, 1123 (1974).
57. Felsenfeld, G., Miles, H. T.: Biochem. **36**, 407 (1967).
58. Giacomoni, D., Spiegelman, S.: Science **138**, 1328 (1972).
59. Billeter, M. A., Weissman, C., Warner, R. C.: J. Mol. Biol. **17**, 145 (1966).
60. Adams, A., Lindahl, T., Fresco, J. R.: Proc. Nat. Acad. Sci. USA **57**, 1084, (1967).
61. Hess, E. L., Horn, R.: J. Mol. Biol. **10**, 541 (1964).
62. Utsunomiya, T., Roth, J. S.: J. Cell Biol. **29**, 395 (1966).
63. Artman, D., Silman, N., Englesberg, H.: Biochem. J. **104**, 878 (1967).
64. Robertson, H. D., Webster, R. E., Zinder, N. D.: J. Biol. Chem. **243**, 82 (1968).
65. Spahr, P. F.: J. Biol. Chem. **239**, 3716 (1964).
66. Danner, J., Morgan, R. S.: Biochim. Biophys. Acta **76**, 652 (1963).
67. Roth, J. S.: Biochim. Biophys. Acta **61**, 903 (1962).
68. Girija, N. S., Sreenivasan, A.: Biochem. J. **98**, 562 (1966).
69. Grief, R. L., Eich, E. F.: Biochim. Biophys. Acta **286**, 350 (1972).

70. Smeaton, J. R., Elliot, W. H.: Biochim. Biophys. Acta **145**, 547 (1967).
71. Horikoshi, K.: Biochim. Biophys. Acta **240**, 532 (1971).
72. Apirion, D.: Molec. Gen. Genet. **122**, 313 (1973).
73. Bluhm, L., Ordal, Z. J.: J. Bacteriol. **97**, 140 (1969).
74. Rosenthal, L. J., Martin, S. E., Pariza, M. W., Iandolo, J. J.: J. Bacteriol. **109**, 243 (1972).
75. Clark, C. W., Witter, L. D., Ordal, Z. J.: Appl. Microbiol. **16**, 1764 (1968).
76. Hurst, A., Hughes, A., Beare-Rogers, J. L., Collins-Thompson, D. L.: J. Bacteriol. **116**, 901 (1973).
77. Tomlins, R. I., Ordal, Z. J.: J. Bacteriol. **107**, 134 (1971).
78. Miller, L. L., Ordal, Z. J.: Appl. Microbiol. **24**, 878 (1972).
79. Mukherjee, P., Bhattacherjee, S. B.: J. Gen. Microbiol. **60**, 233 (1970).
80. Allwood, M. C., Russell, A. D.: Adv. Appl. Microbiol. **12**, 88 (1970).
81. Wood, T. H.: Adv. Biol. Med. Phys. **4**, 119 (1956).
82. Zamenhof, S., Greer, S.: Nature **182**, 611 (1958).
83. Zamenhof, S.: Proc. Nat. Acad. Sci. USA **46**, 101 (1960).
84. Schenberg-Frascino, A., Moustacchi, E.: Molec. Gen. Genet. **115**, 243 (1972).
85. Schenberg-Frascino, A.: Molec. Gen. Genet. **117**, 239 (1972).
86. Bridges, B. A., Ashwood-Smith, M. J., Munson, R. J.: Biochem. Biophys. Res. Comm. **35**, 193 (1969a).
87. Bridges, B. A., Ashwood-Smith, M. J., Munson, R. J.: J. Gen. Microbiol. **58**, 115 (1969b).
88. Pauling, C., Beck, L. A.: J. Gen. Microbiol. **87**, 181 (1975).
89. Matsumoto, S., Kagami-Ishi, Y.: Jap. J. Genet. **45**, 153 (1970).
90. Macphee, D. G.: J. Gen. Microbiol. **76**, 441 (1973).
91. Haynes, R. H.: Photochem. Photobiol. **3**, 429, (1964).
92. McGrath, R. A., Williams, R. W.: Nature **212**, 534 (1966).
93. Eigner, J., Boedtker, H., Michaels, G.: Biochim. Biophys. Acta **51**, 165 (1961).
94. Greer, S., Zamenhof, S. J.: J. Mol. Biol. **4**, 123 (1962).
95. Dugle, D. L., Dugle, J. R.: Can. J. Microbiol. **17**, 575 (1971).
96. Woodcock, E., Grigg, G. W.: Nature (New Biol.) **237**, 76 (1972).
97. Sedgwick, S. G., Bridges, B. A.: J. Gen. Microbiol. **71**, 191 (1972).
98. DeLucia, P., Cairns, J.: Nature **224**, 1164 (1969).
99. Gomez, R. F., Sinskey, A. J., Davies, R., Labuza, T. P.: J. Gen. Microbiol. **74**, 267 (1973).
100. Krakoff, I. H., Brown, N. C., Reichard, P.: Can. Res. **28**, 1559 (1968).
101. Sinha, N. K., Snustad, D. P.: J. Bacteriol. **112**, 1321 (1972).
102. Gomez, R. F., Blais, K., Herrero, A., Sinskey, A. J.: J. Gen. Microbiol. **97**, 19 (1976).
103. Ramareddy, G., Reiter, H.: J. Bacteriol. **100**, 724 (1969).
104. Goulian, M.: Ann. Rev. Biochem. **40**, 855 (1971).
105. Gomez, R. F., Sinskey, A. J.: J. Bacteriol. **115**, 522 (1973).
106. Ganesan, A. K., Smith, K. C.: Cold Spring Harbor Symp. Quant. Biol. **33**, 235 (1968).
107. Gomez, R. F., Sinskey, A. J.: J. Bacteriol. **122** 106 (1975).
108. Rosenkranz, H. S., Carr, H. S., Morgan, C.: Biochem. Biophys. Res. Comm. **44**, 546 (1971).
109. Murthy, M. S. S., Rao, B. S., Reddy, N. M. S., Subrahmanyam, P., Madhvanath, U.: Mut. Res. **27**, 219 (1975).
110. Bockrath, R., Cheung, M. K.: Mut. Res. **19**, 23 (1973).
111. Van der Schueren, E., Youngs, D. A., Smith, K. C.: Photochem. and Photobiol. **20**, 9 (1974).
112. Sedgwick, S. G.: J. Bacteriol. **123**, 154 (1975).
113. Harris, W. J.: Biochem. Soc. Trans., Dublin **1**, 240 (1973).
114. Inman, R. B.: J. Mol. Biol. **28**, 103 (1967).
115. Ioannou, P.: Nature New Biol. **244**, 257 (1973).
116. Beerman, T. A., Lebowitz, J.: J. Mol. Biol. **74**, 451 (1973).
117. Chamberlin, M., Berg, P.: Cold Spring Harbor Symp. Quant. Biol. **28**, 67 (1963).
118. Saucier, J. M., Wang, J. C.: Nature New Biol. **239**, 167 (1972).
119. Goldmark, D. J., Linn, S.: Proc. Nat. Acad. Sci. USA **67**, 434 (1970).
120. Verly, W. G., Paquette, Y., Thibodeau, L.: Nature (New Biol.) **244**, 67 (1973).

Cellular and Microbial Models in the Investigation of Mammalian Metabolism of Xenobiotics

R. V. SMITH and D. ACOSTA, JR.
Drug Dynamics Institute and Department of Pharmacology, College of Pharmacy, University of Texas At Austin, Austin, Texas 78712 U.S.A.

J. P. ROSAZZA
Division of Medicinal Chemistry and Natural Products, College of Pharmacy, University of Iowa, Iowa City, Iowa 52242 U.S.A.

Contents

1. Introduction

In advanced technological societies, man and other mammals are being challenged by an ever increasing number and variety of foreign organic compounds (xenobiotics). Whether unintentially consumed (environmental pollutants) or purposefully employed for some beneficial effect (drugs), the biological effects of such agents is frequently predicated on their biotransformations in mammals. That is, metabolites formed *in vivo* may be responsible for some or all of the biological effects of chemicals; alternatively, detoxification and/or enhanced elimination of foreign substances may be influenced by metabolic transformations. Because of its importance in understanding the biological effects of foreign substances, there is increasing interest in defining systems for study of the metabolism of xenobiotics.

Traditionally, drug metabolism studies are implemented with live animals. Following administration of foreign substances, excreta are examined for the presence of metabolites. The latter are identified, evaluated for biological activity and an overall assessment made of their importance based on relative distribution to affected organs. This "classical" approach is plagued by several difficulties including:

1) cost of animal specimens;
2) difficult isolation of metabolites from complex biological media;
3) preparation of sufficient amounts of metabolites for biological evaluation.

The first two problems have been tackled by numerous investigators. In most cases, subcellular fractions (principally microsomes) of liver tissue have been utilized as *in vitro* models of mammalian metabolism. However, the frequent disagreement of results observed during parallel *in vivo* and *in vitro* investigations have led many to suspect that enzyme systems may be altered during isolation procedures. Thus, while analytical chemical difficulties are simplified by working with hepatic microsomal systems, the lack of correlation of *in vitro* with *in vivo* results persists. This problem might be solved by employing primary hepatic cell cultures which should be physiologically close in function to cells in intact animals. Furthermore, studies of drug metabolism by cellular systems should simultaneously permit assessment of hepatotoxicity of foreign chemicals. An exploration of the feasibility of this proposal constitutes the second part of this review.

The preparative synthesis of metabolites is a significant problem in drug metabolism studies. Gram-quantities of metabolites are frequently desired for complete structure elucidation and biological testing. While organic synthetic methods provide an obvious solution for metabolites of relatively simple molecules, preparation of metabolites of more complex substances require vast expenditures in time and effort. Recently, Smith and Rosazza [1, 2] described microbial systems that mimic mammalian biotransformations and are potentially useful in the preparative synthesis of drug metabolites. In the first part of this review, methods are outlined for experimentally optimizing conditions for the microbial preparation of difficult-to-synthesize metabolites.

The use of selected microbial systems and primary hepatic cell cultures for study of the metabolism of drugs and other foreign chemicals represent what we refer to as *microbial and cellular models of mammalian metabolism.*

2. Microbial Models of Mammalian Metabolism

In 1952, great impetus was given to the field of microbial chemistry and biochemistry by the momentous report of Peterson and Murray [3] that detailed the use of microorganisms in selectively introducing hydroxyl groups into the steroid nucleus. The 11-hydroxylation reaction accomplished by several microorganisms under mild conditions, gave a difficult-to-synthesize intermediate for the preparation of corticosteroids, and paved the way for further rapid expansion of the field of microbial transformations of steroids. Since 1952, many kinds of substrates have been biotransformed by microorganisms. Some compounds are similar to the steroids (alicyclic and non-aromatic), while others possess much more diverse chemical structures (alkaloids, acids, phenols, alkanes, antibiotics) [4–7]. The possibilities for applying microbial transformations during drug metabolism studies have only been realized in the last few years.

Smith and Rosazza [1, 2] described microbial transformation systems consisting of several microorganisms that collectively mimic many of the kinds of biotransformations observed in mammals. These authors [1, 2] suggested that it is unlikely, although not impossible, that a single microorganism could mimic all of the biotransformations conducted by a single mammalian species. However, a *Cunninghamella* species [8] has been shown to accomplish N- and O-dealkylations, and aromatic hydroxylation including the "NIH shift". Interestingly, cytochrome P-450 oxidizing systems that are recognized as effecting many transformations in mammals, are now known to exist in bacteria [9], yeasts [10, 11] and in filamentous fungi [12].

An important part of the application of microbial models to the study of drug metabolism is the potential for obtaining large quantities of metabolites of drugs *via* fermentation scale-up techniques. The availability of drug metabolites in large quantities enables the direct chemical identification of metabolite structures and permits thorough pharmacological/toxicological study. The need for sometimes lengthy organic synthetic efforts for the preparation of metabolites is thus obviated.

Several reports of large-scale production of drug metabolites may be cited. For example, very high yields of 11-α-hydroxyprogesterone (*2*) were produced by *Aspergillus ochraceous* at high substrate (progesterone, *1*) levels [13]. That is, up to 65% conversions of *1* to *2* were obtained at steroid concentrations of 50 gm/liter, while 90% conversions were observed at progesterone concentrations of 20 gm/liter. Less impressive, but significant, examples of the production of gram quantities of drug metabolites include the hydroxylation of acronycine (*4*) [14], O-demethylations of papaverine (*5*) [15] and the N- and N′-demethylations of *d*-tetrandrine (*6*) [16].

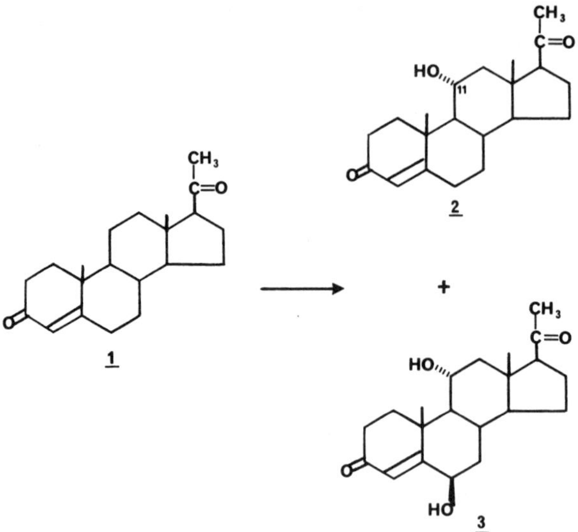

CH3
C=O
HO 11
2
CH3
C=O
1
+
CH3
C=O
HO
HO
3

3. Performing Microbial Transformation Experiments

Methods generally applied in growing microorganisms and in performing fermentation experiments have been well-described elsewhere [1–7, 17, 18]. Although some variation in approach exists, the basic elements of microbial transformation experiments are the same. Considerable emphasis must be given to the control of fermentation parameters [19, 20] since media and other environmental changes can influence the complement of enzymes produced by a microorganism and favor the formation of a single metabolic product. Typical fermentation parameters that are varied include medium composition, pH, aeration of cultures, and temperature.

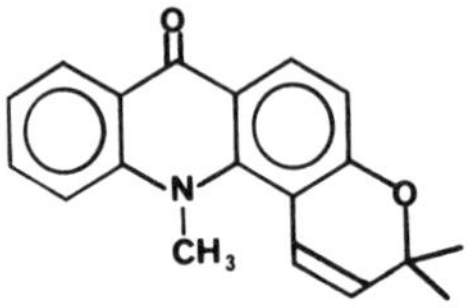

O
N
CH3
O
4 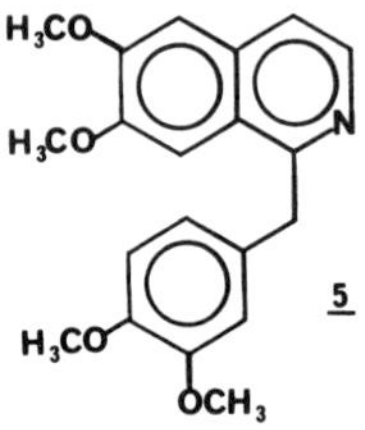
H3CO
H3CO
N
5 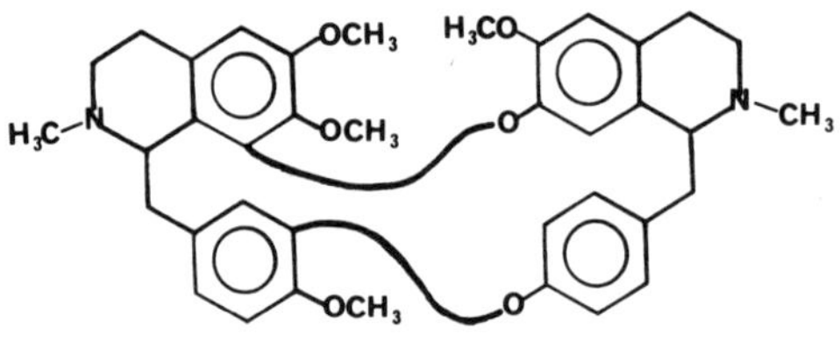
H3CO
OCH3
OCH3
H3CO
H3C–N
OCH3
O
N–CH3
OCH3
O
6

Microorganisms used in microbial transformation studies are obtained from a variety of sources including soil, standard culture collections [5, 7, 18, 21] and sewage treatment facilities [22]. Most experiments are performed with pure culture strains, although several examples may be cited where mixed microbial cultures have been used to achieve biotransformation on substrates like drugs and pesticides [6, 23–27]. It is possible to obtain leads in the literature for cultures which are capable of performing specific types of biotransformations (e.g. O- and N-dealkylations, dehydrogenations, hydrolyses, reductions, or hydroxylations) [1–6, 15–18]. Biotransformations have been reported for practically every type of microorganism including bacteria, yeasts, filamentous fungi and lichens. The choice of medium employed depends on the nature of the organism being grown. Numerous media have been used in microbial transformation experiments [17], and some of these are listed in Table 1. The carbon source chosen, inorganic nutrients and medium pH may drastically influence the outcome of fermentations. For example,

Table 1. A selection of media employed in microbial transformation studies

Ref. [1]

1. Dextrose, 20 gm; yeast extract, 5 gm; soybean meal, 5 gm; NaCl, 5 gm; K_2HPO_4, 5 gm; distilled H_2O, 1 liter; pH 7.0 with 5N HCl (yeasts, fungi, *Actinomycetes*).
2. Dextrose (1 H_2O), 30 gm; ammonium tartrate, 715 gm; KH_2PO_4, 2 gm; $MgSO_4 \cdot 7H_2O$, 0.5 gm; trace elements, 0.1 gm; yeast extract, 1.0 gm; distilled H_2O, 1 liter; pH 5.5 for fungi and streptomycetes; pH 7.0 for bacteria.

Ref. [28]

1. Edamine, 20 gm; cornsteep liquor, 2 gm; Dextrose, 50 gm; H_2O, 1 liter, pH 4.3. Used with *Rhizopus niger.*
2. Peptone, 0.5%; Dextrose, 2%; Soybean meal, 0.5%; KH_2PO_4, 0.5%; yeast extract, 0.5%; NaCl, 0.5%. Used with *Cunninghamella blakesleeana.*

Ref. [29]

1. Inverted blackstrap molasses, 100 gm; cornsteep liquor, 6.3 gm; distilled water, 1 liter
2. Dextrose, 50 gm; lactalbumin digest, 20 gm; cornsteep liquor, 5 gm; distilled H_2O, 1.0 liter. Used with *Aspergillus niger.*
3. Sucrose, 50 gm; $NaNO_3$, 7.6 gm; K_2HPO_4, 1.0 gm; $MgSO_4$, 0.5 gm; KCl, 0.5 gm; $FeSO_4 \cdot 7H_2O$, 1 liter.

both the yields and ratios of 11-α-hydroxyprogesterone (*2*), amd 6-β-11-α-dihydroxyprogesterone (*3*) obtained with *Aspergillus ochraceous* and *Rhizopus nigricans* [28, 29] are significantly effected when fermentations are conducted with different media. Reduction of the 3,4-double bond of *1* by *Nocardia corallina* is markedly influenced by the concentration of glucose in the medium [30a]. A more comprehensive listing of culture media that is cross-indexed with various types of microorganisms is available [30b].
To identify cultures that possess the capacity to metabolize xenobiotics, small-scale screening experiments are typically performed. For screening, cultures are grown in small volumes (10–25 ml) of medium containing xenobiotics at nominal concentrations of 0.1–0.5 mg/ml. Detection of metabolites may be simply accomplished by solvent extraction

and thin-layer chromatographic (TLC) analysis. Although detection procedures are quite simple to perform, considerable effort may be required to insure that they are capable of detecting metabolites even at low amounts (1% conversions). Optimization of the detection system may involve determination of distribution coefficients of substrate and possible metabolites in extraction solvents, the development of crisp and reproducible chromatographic systems, and the selection of spray reagents capable of discriminating drug metabolites. Examples of the TLC systems noted have been developed in our laboratories [31, 32].

A two-stage submerged fermentation process is commonly employed to obtain sufficient cell-mass for conducting biotransformation studies and to insure reproducibility. One-week old slants of microorganisms are flooded with sterile medium (or saline) and gently agitated with an inoculating loop to suspend spores or vegetative growth. In this manner, inocula for shaken-flask cultures are produced. Cotton-plugged Erlenmeyer flasks containing one-fifth of their volumes of sterile medium receive slant inocula and are incubated on rotary shakers operating nominally at 250 rpm and 28° C for two to three days before being transferred to a second flask containing fresh medium. This procedure provides a vigorous, rapidly growing culture as inoculum for the second stage of the fermentation. Commonly, the latter inoculum represents 10% of the volume of medium contained in the second stage culture flask, although inocula volumes may vary considerably (1–10%). Second-stage cultures are incubated, again with shaking for 24 to 30 hrs prior to the addition of drug substrates. Occasionally, it becomes necessary to gently homogenize clumped mycelial pellets in the second-stage culture to provide high cell-surface area which enhances biotransformation yields and insures reproducibility. Substrate-containing cultures are reincubated for various periods of time up to and surpassing one week. These cultures are sampled and assayed at various time intervals, usually by TLC.

In the fermentations described above, growth in the second-stage flask is usually faster and more vigorous than in the primary cultures that are initiated from slants. Some microorganisms do not exhibit a typical growth curve observed with unicellular organisms like bacteria. A few workers have added substrates at zero time to second-stage cultures, but most often, substrates are added after 6 to 48 hrs of second-stage growth. By these times, most second-stage cultures are fully grown, and have nearly consumed nutrients in the culture medium. This can pose an interesting problem since two competing factors may be operating in the microbial transformation experiment that either diminish or increase biotransformation yields. In the presence of large amounts of carbohydrate, for example, an inhibitory "glucose effect", or catabolic repression effect may occur [33–35]. On the other hand, the presence of carbohydrate in the medium may be desirable and necessary since it may be required as a substrate for co-metabolism [34–36].

The scale of the fermentation employed in biotransformation work is dependent on the objectives of the experiment. Screening-scale studies are conducted in small flasks containing 5–25 ml of medium, while preparative scale studies may be conducted in larger Erlenmeyer flasks; in multiliter stirred laboratory fermentors or in larger industrial-scale fermentation equipment. Often it is possible to conduct microbial transformation experiments using resting cells, rather than actively growing cultures (*vide infra*).

Several of the factors which are important in optimizing biotransformation yields have been mentioned above. Additional factors and techniques that can be employed to increase biotransformation yields in drug metabolism studies will be considered below.

3.1 Influence of Physical States of Substrate

It is generally believed that microbial transformations occur with dissolved substrates. Thus, the degree of solubility of a substrate in a bioconversion system effects the yield of a given reaction. One may assume that after dissolution, substrates are carried to the cell wall or are transported within cells to sites where biotransformations occur. Metabolites may pool within or on cell walls or they may partition again to the surrounding medium where they may be detected.

Growth rates of bacteria vary when naphthalene (*7*), phenanthrene (*8*) and anthracene (*9*) are used as carbon sources [37]. The generation times observed with these hydrocarbons (1.5, 10.5 and 29 hrs respectively) reflect their water solubilities [38]. Growth rates of bacteria were found to be indepent of the total amount of solid naphtalene [39] and phenanthrene [40] in a given medium, and were proportional to the amount of dissolved hydrocarbon only. Furthermore, it was found that bacteria used the dissolved hydrocarbons and grew at the interface between the aqueous medium and solid hydrocarbon particles. Generation times were controlled by the rate of substrate dissolution which might or might not match the rate of uptake and oxidation by microbial cells.

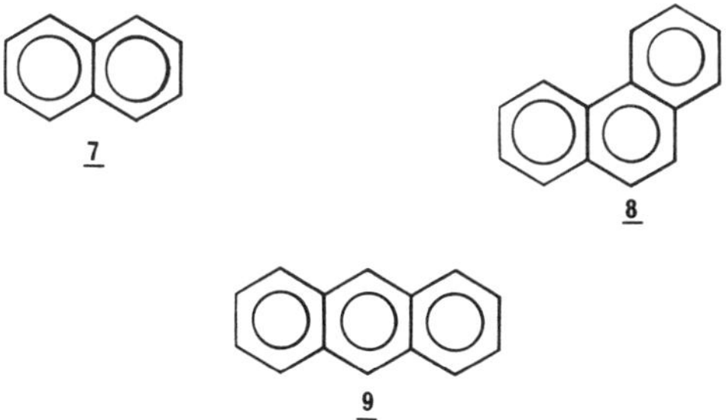

Water soluble substrates are readily dissolved and dispersed in aqueous fermentation media. However, water-insoluble substrates are generally added to fermentations as concentrated solutions in water-miscible solvents such as ethanol, acetone, ethyleneglycol, dimethylsulfoxide, and dimethylformamide. Water-immiscible solvents such as methylene chloride and benzene have also been used to dissolve and administer substrates to fermentations. When concentrated solutions of lipophilic substrates are dispensed into aqueous media, the substrates become very finely divided, often assuming a milk-like appearance. To avoid the potential problem of adding large volumes of toxic solvents to a fermentation, some workers [20] employ a semicontinuous addition process. Karrow and Petsiavas [20] added progesterone to *Aspergillus ochraceous* fermentations in this manner. By using such a "dosing technique", solvent toxicity was minimized, there appeared to be a decrease in the mechanical loss of starting material through aggregation, fewer side reactions other than the desired 11-hydroxylation of progesterone occured, and more accurate timing of the biotransformation reaction was permitted.

Jones and Baskevitch [40] described the influence of aggregation of steroid substrates in aqueous solution on their enzymatic transformations. Kinetic analyses revealed that aggregation of steroids in aqueous media results in a deviation from pseudo-first order kinetic plots under conditions where acid-catalyzed isomerizations occur. Similar analysis has also shown that critical methanol concentrations are required to completely solvate the 3-keto-analogs of stigmasterol (*10*) and β-sitosterol (*11*). The presence of hydroxyl groups on the side chains of the steroids reduce the amount of methanol necessary

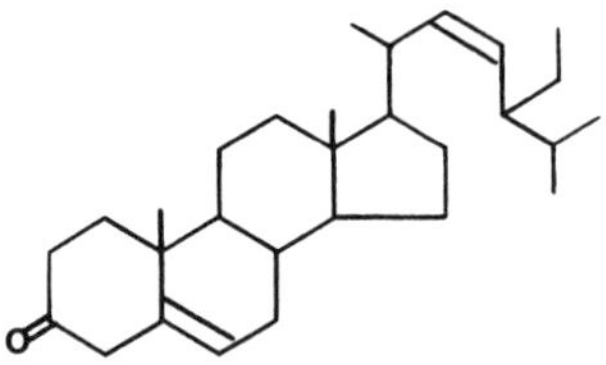

10

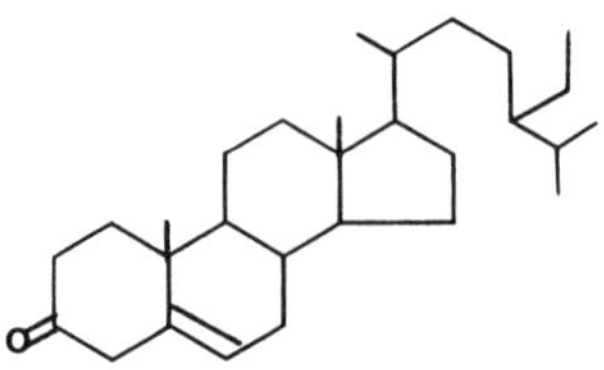

11

to prevent aggregation. Unfortunately, however, the amount of methanol necessary to prevent complete aggregation of compounds like *10* and *11* is in excess of the amount tolerable in most microbial enzymes.

Others have examined the influence of the mode of substrate addition on biotransformation reactions [27]. In mixed cultures, *Arthrobacter simplex* and *Curvularia lunata* accomplish 1-dehydrogenation, and 11-β-hydroxylation of 16-α-hydrocortexolone-16,17 acetonide (*12*). Stock solutions of *12* were prepared as suspensions in cold solvent, or in 0.1% w/v aqueous Tween 80; the substrate was also dissolved in hot or cold solvents. Each suspension or solution was prepared at a concentration of 100 mg/ml. The cultures

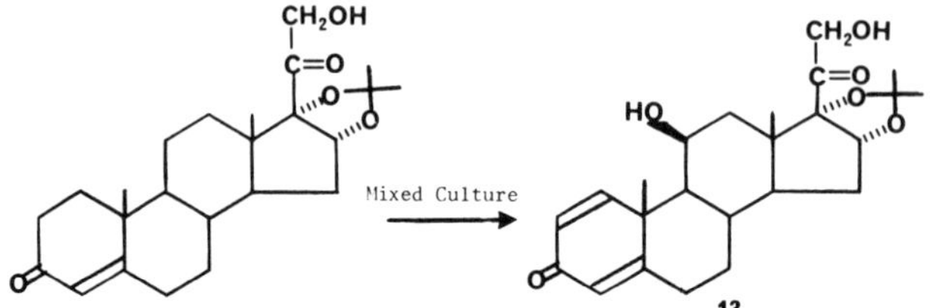

CH_2OH $C{=}O$ 12 Mixed Culture → HO CH_2OH $C{=}O$ 13

were grown separately, and then mixed in a 1 : 5 proportion (*A. simplex : C. Lunata*) using a 6% inoculum. Substrate was added to the mixed culture at zero time (500 μg/ml) and again at 24 hrs (500 μg/ml additional). The results obtained by adding the substrate to the fermentation in different ways are summarized in Table 2. Hot solvents or cold DMF caused immediate precipitation of *12* when added to the medium, and the precipitate was different in crystal form and size than the original steroid; particle size of the precipitated steroid were determined by a Coulter Counter. Hot ethanol and hot ethanol-boric acid-water solutions yielded particles 0.5–2 μm in size, while cold steroid

Table 2. The effect of the mode of addition on the conversion of *12* to *13* [27]

Solvent	% Conversion to *13*
Hot Borate-Ethanol-Water[a]	90%
Hot Ethanol	65%
Hot Ethanol-Acetone	67%
Cold Dimethylformamide	60%
Cold Ethanol-Acetone Suspension	30%
Cold Ethanol Suspension	20%
Cold Aqueous 0.1% Tween 80	10%

[a] As a 1% solution in sodium tetraborate (0.15%)-ethanol (94%).

suspensions gave particles ranging from 10–100 μm in size. The growth of *C. lunata* was also influenced by the solvents used, ranging from filamentous to granular or pelleted mycelial aggregates. No changes were observed in the growth of *A. simplex*.

Conclusions derived from this study indicate that relatively rapid rates of steroid bioconversions are possible when substrates are added as solutions, not as suspensions. The ultrafine particles formed in the media appeared to be amorphous. It is well-known that polymorphic forms of various compounds differ in solubility and amorphous forms are generally more soluble than crystalline forms. The effect of sodium tetraborate is believed to be due to changes in the permeability of cells to the substrate. Consistent with the findings of Jones and Baskevitch [40], larger particle sizes, or factors which cause substrate aggregation may decrease biotransformation yields.

The actual crystallization of a product from a fermentation has been reported [41]. During the δ-1-dehydrogenation of hydrocortisone to prednisolone by *C. simplex*, the product precipitated out, forcing the reaction to near completion.

Zeolites have been used to enhance the solubility of hydrocarbons in aqueous media. Hydrocarbons are physically entrapped within and over the large surface area of the zeolites; growth rates of microbial cells surrounding or growing into hydrocarbon-containing particles were observed to increase [42]. This technique does not appear to have been applied in microbial transformations of low molecular weight drug substances.

Weaver *et al.* [13, 43] employed a milling and wetting technique to improve yields when large amounts of progesterone (*1*) were added to cultures of *Rhizopus nigricans* and *Aspergillus ochraceous.* Progesterone was used without treatment or by passing it through a "Jet-o-Mizer" grinder to yield a form approximately one-third as dense as the original compound. The milled steroid was wetted with 0.01% Tween 80 and sterilized with live steam before being added to microorganisms. Using this technique, *Rhizopus niger* was capable of carrying out the 11-α-hydroxylation of *1* in amounts of up to 8 gm/liter in 50% yield. *A. ochraceous* metabolized up to 20 gm of *1*/liter almost quantitatively while 65% conversions were obtained at substrate levels of 50 gm/liter. Yields with milled progesterone were nearly 50% better than those obtained with untreated progesterone. Tween 80 prevented the steroid from floating to the surface in aqueous medium. In these high yielding incubations, substrates and products were isolated by simple filtration. The high concentration of substrates in these incubations prevented side reactions from occurring; particularly, the formation of 6-β-11-α-dihydroxyprogesterone (*3*). Side reactions were also avoided by adding the steroid to the medium in a semi-continuous manner.

3.2 Transformations in Nonaqueous Media

Since microbial transformations occur most readily with soluble substrates, it is logical to suggest that such reactions might be conducted successfully in nonaqueous solvents, or in mixtures of aqueous and nonaqueous media. Such a proposal has several advantages. The usual microbial transformation experiment using insoluble substrates requires large volumes of medium. Use of nonaqueous solvents capable of dissolving the substrate could obviate this problem. In two-phase systems (emulsions), enzymes or cells would be protected by an aqueous environment, while lipophilic substrates and potentially inhibitory products would tend to remain in the non-aqueous phase. Rapid reaction rates should be realized if the solutes (completely solubilized one phase) rapidly transfer to the other phase for biotransformation. Dissolved oxygen levels are considerably higher in nonaqueous solvents than in aqueous media. Disadvantages of nonaqueous solvents for biotransformation experiments include the prohibitive use of volatile and flammable solvents. Additionally, polar nonaqueous solvents tend to dehydrate and denature microbial enzymes.

Several recent reports detail the use of enzymes in nonaqueous systems which may serve as precedent for similar experiments with intact microorganisms. Cooney and Hueter [44] described studies of a chloroperoxidase enzyme from *Caldariomyces fumago*. Experiments with solutions of substrate in 10% dimethylsulfoxide were successful but only 33–47% of the full enzyme activity was observed. Trypsin attached to a synthetic carrier [45] demonstrated activity in 50% dimethylformamide solution.

Cremonesi *et. al.* [46–49] employed emulsions of buffer with various organic solvents to conduct enzymatic bioconversions of a number of steroid substrates. A range of solvents were tried to determine the activity of fungal laccases (copper-containing oxidases) from *Polyporus versicolor* with estradiol [49]. Ether, butyl acetate, ethyl acetate and methyl ethyl ketone served as the best solvents, primarily because they were capable of dissolving the steroid; more polar solvents caused the inactivation of the enzyme. In the incubations noted, the enzyme was dissolved in the aqueous phase while the substrates and products were contained in the nonaqueous phase.

The enzymatic dehydrogenation of testosterone (*14*) in a two-phase system has been described [48]. The reaction was enhanced when the β-hydroxysteroid dehydrogenase from *Pseudomonas testosteroni* was coupled to lactic acid dehydrogenase (LDH) from rabbit muscle [46] (Fig. 1). The yields of 4-androstene-3,17-dione (*15*) were directly dependent on the amount of pyruvic acid (*16*) added and its concurrent conversion to lactic acid (*17*). In the coupled reactions, both the steroid dehydrogenase and LDH remained in the aqueous phase while the steroid substrate primarily remained in the nonaqueous phase. Incubations were conducted in 3 ml of buffer containing NAD^+, LDH, and β-hydroxysteroid dehydrogenase. This mixture was added to 3 ml of organic solvent containing steroid, and the two phases were intimately mixed by shaking at 90

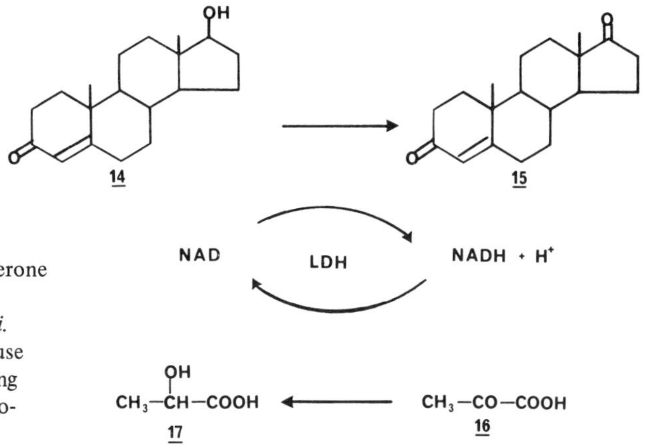

Fig. 1. Dehydrogenation of testosterone (*14*) with *beta*-hydroxysteroid dehydrogenase from *P. testosteroni.* Per cent conversions enhanced by use of nonaqueous solvents and coupling to rabbit muscle lactic acid dehydrogenase [48]

oscillations per minute. Steroid conversions were best in butyl acetate (100% yield). Lower yields were obtained with ethyl acetate (90%) and butanol (15%). A ratio of greater than 3 : 1 of LDH : β-hydroxysteroid dehydrogenase was required for optimum enzymatic activity.

Cremonesi *et al.* [47] have also coupled a 20-β-hydroxysteroid dehydrogenase system with alcohol dehydrogenase (ADH) and a chemical couple (semicarbazide) to force the reaction to completion (see Fig. 2). Semicarbazide is required, since the equilibrium constant of the overall reaction without the reagent is less than 1 at 25° C. Incubations were conducted in equal volumes of nonaqueous solvent and buffer containing the steroid dehydrogenase (usually stabilized by addition of serum albumin), alcohol dehydrogenase, ethanol and semicarbazide. As with other enzymes, the stability of the dehydrogenase decreased with increasing solvent polarities. Solvents less polar than diethyl ether gave the highest rates of conversion. Butyl acetate gave 100% yields, while other solvents provided yields in the range of 5–15%. In a preparative scale conversion, 2.77 mmole of cortisone (1.0 gm) dissolved in 500 ml of butyl acetate were reacted

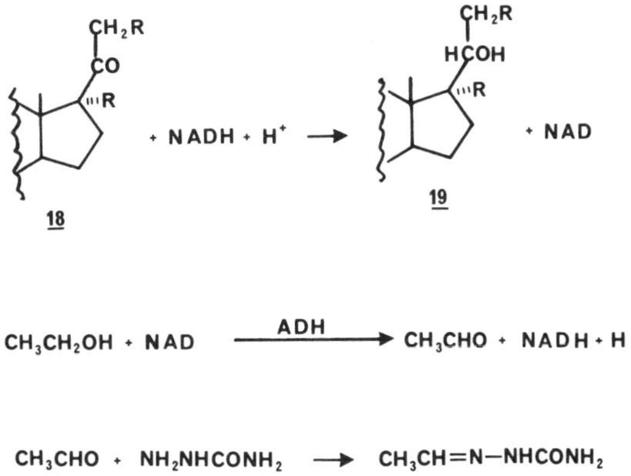

Fig. 2. Reduction of 20-ketosteroids using a hydroxysteroid dehydrogenase coupled with an alcohol dehydrogenase. Nonaqueous solvents and a semicarbazide chemical-couple were used to improve yields [47]

with 500 ml of 0.05 M phosphate buffer containing 29 units of 20-β-hydroxysteroid dehydrogenase, 5 mg ADH, 0.125 mole NAD, 15 ml ethanol, 2.98 mmole of semicarbazide and 150 mg of serum albumin. The reaction was complete in 4 hrs. Other steroids reduced in this manner included prednisone, cortisol, deoxycorticosterone and prednisolone. The authors [47] emphasized that chemical reductions in such systems are much less efficient.

Whole resting cells of *Nocardia* species were employed to convert cholesterol (*20*) into cholest-4-ene-3-one (*21*) in a nonaqueous solvent mixture [50].

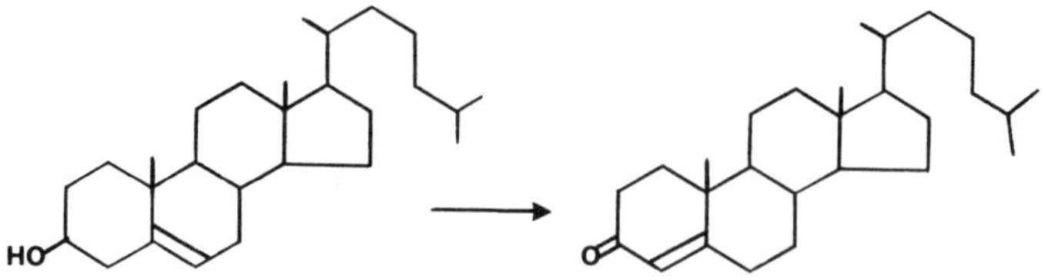

HO

20

21

Nocardia (NCIB 10554) was grown in a 1000 L fermentor where dissolved oxygen was maintained between 30 and 50% saturation during growth. Cholesterol oxidase was induced with cholesterol in the medium. *Nocardia* cells were harvested by centrifugation prior to growth cessation and were stored at –20° C until required for use.

The conversion of *20* to *21* in various solvents was determined by reacting 2 ml of organic solvent containing 1% w/v of cholesterol with 0.1 gm wet weight of *Nocardia* cells. Densitometric or spectrophotometric analyses were used to monitor yields listed in Table 3.

In the *Nocardia* experiments, described above, thawed cells showed rates of reaction nearly independent of the ratio of aqueous/organic solvents. However, freeze-dried cells

Table 3. Conversion of cholesterol (*20*) to cholestenone (*21*) by *Nocardia* resting cells in various solvents [50]

Solvent	% Conversion	O_2 Solubility (mM)
Toluene	24	7.5
Hexadecane	22	5.1
Carbon tetrachloride	21	10.3
Ether	18	18.5
Benzene	15	7.3
Pet ether (60–80° C)	10	18.3
Cyclohexane	8	–
Chloroform	8	9.15
Acetone	3	9.25
Isopropanol	3	–
Ethanol	2	6.38
Water	3	1.26

were inactive, in the absence of aqueous buffer. Rates of reaction were generally reproducible and most rapid in carbon tetrachloride solutions. When an incubation of 2 gm of cells in 6 ml of 50 mM pH 7.0 phosphate buffer and 40 ml of carbon tetrachloride was performed, the rate of conversion obtained was 1.25 gm/l/h. The reaction rates were proportional to the amounts of cells used and to the rate of stirring. They were also influenced by dissolved oxygen levels and temperature. *Nocardia* cells used in a single experiment lost 45% of their cholesterol oxidase activity when stored overnight at 4° C. Frozen cells on the other hand, were less stable. In one experiment, after 69 hrs of use and seven runs, the conversion rate of the same cells dropped to 52% of their original activity [50].

3.3 Applications of Resting Cell Suspensions

Biotransformation experiments are conducted most often with actively growing cultures submerged in fermentation media. The complexity of such media may render the processes of isolation and subsequent purification of microbial metabolites quite difficult. Lipids, phenolic compounds, carboxylic acids and other kinds of normal microbial metabolites may have similar physical properties to desired metabolites, and may be coextracted. As a result, the demands on chromatographic purification systems can be significantly increased. Resting cell suspensions of microorganisms can serve as an alternative to performing transformations with growing cells in nutrient media. Microorganisms may be grown to a desired stage of growth, harvested by filtration (fungi), or centrifugation (bacteria, yeasts), and resuspended in buffer or in buffers containing small amounts of glucose or other nutrients. Resting cells offer several advantages over the use of growing cultures and isolated enzymes in biotransformation processes, including: 1) accomplishment of multistep enzymatic conversions without the use of coenzymes; 2) increased efficiency of organized enzymes as found in intact microorganisms; 3) avoidance of losses in enzyme activity due to isolation procedures; 4) enhanced stability compared to isolated enzymes; 5) little possibility of microbial contamination; 6) simplified metabolite isolations. Resting cell suspensions may be disadvantageous if cooxidation substrates (glucose or other carbon sources) are required to supply energy for desired biotransformations or if metabolites pool within or on cell walls making product isolations difficult.

As noted earlier, a *Nocardia* species has been employed in the oxidation of cholesterol [50] in resting cell suspensions using organic solvents. The ability of resting cell suspensions of *Septomyxa affinis* (ATCC 13425) to accomplish δ-1-dehydrogenation of a variety of steroids has been described [51]. The dehydrogenase was induced by addition of progesterone or 3-keto-bisnor-4-cholene-22-1 to growth media.

Cunninghamella bainieri (ATCC 9244) was used to transform the anthelmintic agent, methyl-5(6)-butyl-2-benzimidazolecarbamate (*22*) into two of its mammalian metabolites, *23*, and *24* [52]. Small scale incubations were conducted in 50 ml of buffer or broth. Larger scale incubations were done by incubating *C. bainieri* in several liters of medium. Cultures were grown at 200 rpm and 30° C while air was sparged in at a rate of 1 volume/volume-medium/minute for 48 hrs. Cells were isolated by centrifugation and resuspended in 9.5 liters of 1.5% sterile glucose solution to which 5 gm of *22* was added.

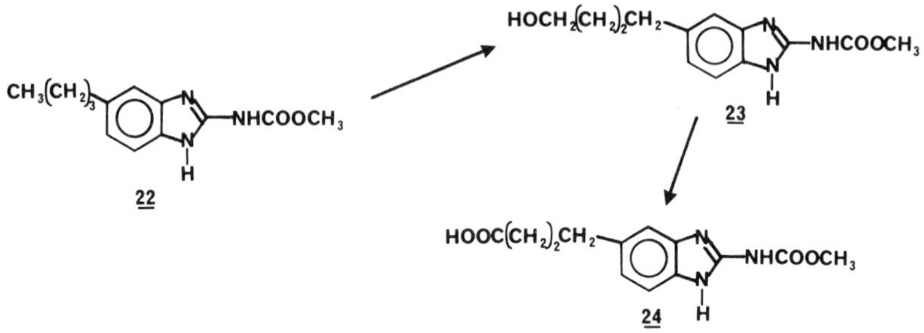

$CH_3(CH_2)_3$–[benzimidazole]–$NHCOOCH_3$ (22) → $HOCH_2(CH_2)_2CH_2$–[benzimidazole]–$NHCOOCH_3$ (23) → $HOOC(CH_2)_2CH_2$–[benzimidazole]–$NHCOOCH_3$ (24)

Resting cells of *Sepedonium ampullosporum* were used in pilot plant scale to perform 16-α-hydroxylation reactions on steroidal substrates [51]. The organism was grown without an inducer in a multistage fermentation in volumes of 10, 250, and 1400 L. The mycelium was isolated by filtration at room temperature over diatomeceous earth filter cakes, rotary drum vacuum filtration, or centrifugation in basket centrifuges. The resulting cells were resuspended in distilled water or phosphate buffer at pH 7.0 within 6 hrs of the beginning of harvest. Incubations were conducted by resuspending cells without washing in tanks or jars at a concentration of 50 gm of packed wet cell cake/liter (5–6 gm/L dry weight of cells), and aeration, temperature and agitation conditions were the same as those used during growth. After 1 hr, substrates (such as *25*) were added at concentrations of 0.25 gm/L in methylene chloride. By using resting cell suspensions, conversion times were decreased relative to those observed in whole cultures, principally due to differences in the density of cells used. Furthermore, the suspended cells were found to be more stable than cells in the original fermentation broth. Up to 12 sequential batches of 0.25 gm/L each could be converted completely by a single batch of cells. A significant variable in resting cells was the dissolved oxygen level. Oxygen tensions of greater than 90% saturation were required for maximum conversion rates. Several other examples of resting cell suspension conversions can be cited [53–59].

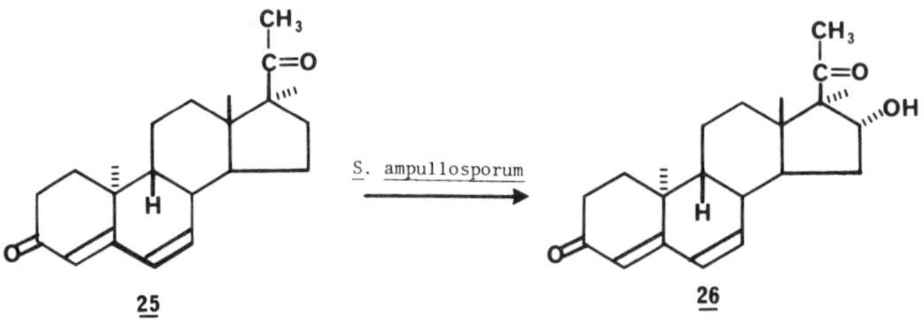

25 —*S. ampullosporum*→ 26

An additional novel application of resting cells has been described for the microbial oxidation of steroids [60]. *Rhizopus nigricans* grow in globular mycelial pellets which can be packed into a column through which nutrient broth is circulated in a continuous manner. At time intervals, progesterone (*1*) was added, while 11 α-hydroxyprogesterone

(*2*) was produced. The hydroxylated product was found in the fermentation beer, while the unreacted starting materials could be recovered from the cells. This application is interesting since other kinds of cultures assume globular or pelleted growth including the various *Cunninghamella* species [1, 2, 14, 52].

3.4 Use of Polymer-Entrapped Cells

Immobilization techniques have been widely applied in modern enzyme technology where isolated enzymes are bound onto or trapped within polymeric matrices [61–63]. Immobilized enzymes have been used in a variety of situations to accomplish selective biotransformations, often in continuous column processes where solutions of substrates are caused to pass over the enzyme and products appear in column effluents. Since enzyme stability in immobilized reactors is often superior to that observed in solution, it has been suggested that enzymes, as they occur in cells, are compartmentalized or entrapped into matrices which may account for natural stability.

The instability of some kinds of purified enzymes renders them impractical for immobilization and application in bioreactors; purification of enzymes often represents a time-consuming and expensive proposition. Furthermore, most enzymes permit only single transformations with a given substrate. For these reasons, several investigators have applied immobilization technology to the entrapment of microbial resting cells. Entrapped microorganisms have been used to conduct specific biotransformations, and since they are in the resting state, many of the advantages and disadvantages of using resting cells in biotransformation experiments also apply to polymer-entrapped microbial cells. It appears that one major advantage in the use of polymer-entrapment techniques is the ease with which cells may be handled in columns or other biochemical reactors.

Sticks and Updike [64] have described detailed methodology for the preparation of lyophilized polyacrylamide enzyme gels for chemical analysis, and their technique has been used in the polyacrylamide gel entrapment of microbial cells. Glucose oxidase and lactic acid dehydrogenase were entrained in a gel which was prepared through photopolymerization employing acrylamide monomer, N,N-methylene-bis-acrylamide crosslinking reagent, riboflavin and potassium persulfate. Nitrogen was bubbled into the photopolymerization reaction to prevent inhibition of polymer formation by oxygen while heat generated during polymerization was dissipated by placing reaction vessels in an ice bath. The polymerized enzyme gel was sieved, washed with phosphate buffer, lyophilized, and stored under refrigeration. The stabilities of the enzyme were high; lyophilized glucose oxidase showed no loss of activity on storage for over three months, while the LDH gel lost only 10% of the activity per month over a 10 month period. Hydrated enzyme preparations of glucose oxidase lost less than 5% of its activity in 6 weeks at 5° C, while LDH lost most of its activity in 3 months at that temperature. Wieland used the above-mentioned technique in the entrapment of lactic acid dehydrogenase (LDH), alcohol dehydrogenase (ADH) from yeast, and trypsin [65]. Leuschner described an interesting process for entrapment of glutamate-pyruvatetransaminase in cellulose nitrate [66]. This was accomplished by suspending the enzyme in a solution of cellulose nitrate which was then poured onto glass plates and dried. Membranes were then soaked in water, cut into "lamellae" and added to reaction mixtures in large-scale reactors.

Mosbach and Mosbach [67] employed cross-linked polyacrylamide for the entrapment of orsellinic acid decarboxylase, trypsin, and lichen cells containing decarboxylase and esterase. An advantage in employing the lichen cell entrapment process was that enzymes with normally transient existence were able to be used for relatively long periods of time. Mosbach and Larsson [68] employed this same technique in the entrapment of fungal cells of *Curvularia lunata* and the δ-1-dehydrogenase from *Corynebacterium simplex,* both of which were used in the modification of steroids. The gel-entrapped fungal cells were applied to the microbial transformation of Reichstein compound S (*27*) leading to cortisol (*28*) through an 11-β-hydroxylation step. The partially purified dehydrogenase from *C. simplex* was used to convert cortisol (*28*) to prednisolone (*29*). Gel-cell granules could be reactivated after use by suspension in a solution containing cornsteep liquor, sucrose, sodium chloride, Tween 80, and 0.01% Reichstein substance S. By shaking this mixture for 16 hrs at 28° C, the resulting reactivated cells gave four times the activity of nonactivated cells. Thus, they could be reused. The dehydrogenase could be packed into a column and run as such while conversion abilities were related to flow rates through the column.

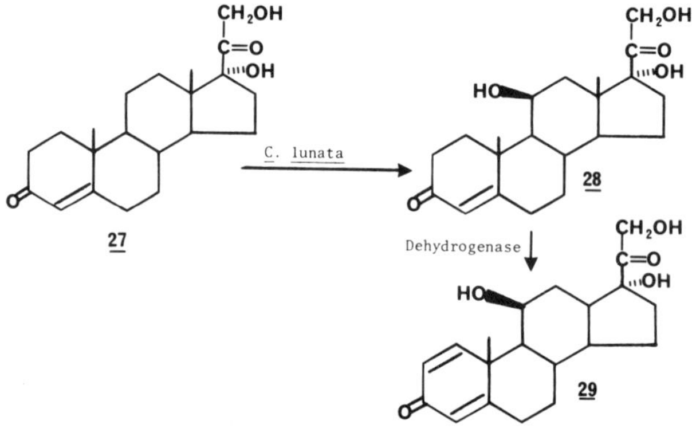

Streptomyces phaeochromogenes cells containing glucose isomerase were entrapped in a collagen matrix [69]. Cells were grown, heated at 80° C for 1.5 hrs and cooled to room temperature before mixing with a pH 6.5 collagen solution. Ratios of cells/collagen were 1/5 to 1/3. After initial flocculation was observed, the pH of the aggregated cell-collagen mixtures was changed to 11.2 with sodium hydroxide. Collagen microfibrils and cells were re-dispersed, and the well-dispersed mixture was cast on a Mylar sheet at a thickness of 2–1 mm. The resulting membrane was tanned by immersion in 10% alkaline formaldehyde or glutaraldehyde solution at pH 8 for 0.5 to 5 min. This "tanning process" gave mechanical strength to the gel. After cutting into small chips, the collagen gel was used batchwise in a reactor. Collagen supports the cell masses in such a way that it resembles "tissues" where the cells are bound by numerous physicochemical bonds to collagen. A near-membrane cell collagen matrix forms which is presumably very stable. Similar techniques have also been used in work with primary hepatic cells (*vide infra*).

Glucose isomerase-containing cells (species undefined) have also been entrained into cellulose acetates [62]. The whole cells are typically entrapped by slurrying in methylene chloride solutions of cellulose acetates. The solvent is evaporated by casting the mixture on a flat surface or by coagulating the mixture in fiber form by passage through a syringe needle. Fibers measuring 250 x 500 μ in cross section and membranes of 10–20 μ thick are produced. The use of cellulose acetate as an entrapping method for whole cells is inexpensive and does not require specific covalent binding methods used for enzymes. The operation is mild, simple and since cellulose acetates are used for artificial kidneys, and for packaging materials, they may find acceptable use in both the food and drug industry.

Pseudomonas putida (ATCC 4359) cells immobilized in a polyacrylamide gel lattice have been employed in the continuous production of L-citrulline. These cells contain L-arginine deaminase [70] required for the conversion. Franks described the polyacrylamide entrapment of *Streptomyces faecalis* (ATCC 8043) cells which were used in the catabolism of L-arginine [71]. In this case, a multistep enzymatic conversion may be obtained without the addition of required coenzymes, thus avoiding the necessity of isolating each of the enzymes involved in the sequence. *Escherichia coli* (ATCC 11303) cells have also been trapped in polyacrylamide matrices [72] by blending in a Waring blender. After activation with ammonium fumarate at 37° C, for 48 hrs, L-aspartic acid could be continuously produced by passing the former through a column packed with the gel. The flow rates of the ammonium fumarate substrate solutions through the column influenced the yield of aspartic acid obtained. Tosa *et al.* [72] also employed polyacrylamide gels to entrap *Bacterium succinium* IAM 1017, *Proteus vulgaris* (OUT 8226), *Pseudomonas aeruginosa* (OUT 8252), and *Serratia marcescens* (OUT 8259) [73], all of which are aspartase-containing microbial cells. Microbial cells have also been used in clumns of some types of Sephadex gels [34].

3.5 Ion Exchange and Dialysis Techniques to Remove Inhibitory Products

The microbial transformation of many kinds of substrates has been documented [4–7]. Often, as with the steroids, large amounts of substrates may be converted to products without any deleterious effect on the culture performing the biotransformation. However, potentially toxic metabolites may be produced which affect either the growth of the microorganism, or one or more of the enzyme systems required for a given biotransformation. In cases such as these, fermentation transformation yields may be enhanced by removing inhibitory products from culture media. Ion exchange resins and dialysis fermentation techniques have been successfully employed for this purpose.

Schwartz and Margalith utilized Dowex anion exchangers to enhace the yields of certain nucleotides produced by *Streptomycetes* [75]. Under ordinary fermentation conditions, considerable amounts of inosine-5′-phosphate (IMP) (*30*), and xanthosine-5′-phosphate (XMP) (*31*), but only traces of guanosine-5′-phosphate (GMP) (*32*) are produced. The

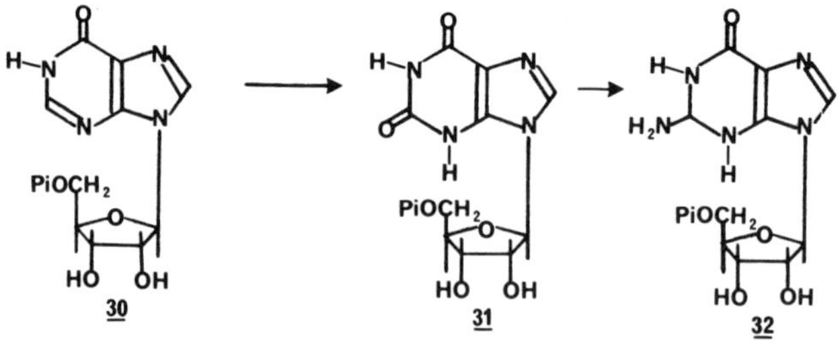

30 → 31 → 32

biosynthetic pathway of these nucleotides is under control of the end product, GMP, which prevents excessive necleotide synthesis through feedback inhibition of the first enzyme in the pathway, IMP-dehydrogenase. Attempts were made to remove the inhibitory GMP from fermentations by complexation with dioxane, and with anion-exchange resins. The resin used in the fermentation was initially treated with sodium hydroxide, washed thoroughly with distilled water and then equilibrated with salt solutions having a similar composition to the fermentation medium. The latter treatment was proposed as a means of avoiding an imbalance of organic ions when the ion exchange resin was added to the medium. The resion was dried in a vacuum oven, sterilized, suspended in sterile distilled water and aseptically added to the culture media 48 hrs after the start of a fermentation. No changes in mycelial growth or pH were observed when the resin was used, and the amount of GMP produced was doubled. The resin also facilitated the isolation step since it bound 95% of the nucleotides produced.

Tone *et al.* [76, 77] employed an Amberlite anion exchanger to trap salicylic acid produced during the degradation of naphthalene by *Pseudomonas aeruginosa* (B 344). The reson was prepared in a manner similar to the Dowex ion exchanger noted above and was equilibrated with medium to prevent problems due to pH change and nutrient absorption. The treated resin was used either directly or packed in cellophane tubes; after autoclaving, the resin was added to cultures 20 hrs after seeding. At the end of a fermentation experiment the resin was recovered by filtration, washed with water, and then treated with hydrochloric acid and ethanol which effected a recovery of 95% or more of the salicylic acid produced. Yields of salicylic acid obtained when 1.0-gram quantities of naphthalene were incubated with the microorganism under various conditions were: without the ion-exchange resin, 14%; with ion-exchange resin, 52%; with ion exchange resin packaged in cellophane, 76%. It was suggested [76, 77] that the dialysis tubing prevented undesirable contact of the resin with microbial cells accounting for the observed high yield. Raymond *et al.* [78, 79] applied weak or strong anion exchangers in the removal of carboxylic acids produced from toluene, xylene and naphthalene by *Nocardia* species. Styrene-divinylbenzene ion exchangers were used in this work, and the carboxylic acid products were primarily recovered from the resins. For example, *p*-xylene is converted into *p*-toluic acid (*34*), and a dihydroxylated analog (*35*) in a ratio of 1 : 3 when inculated with *Nocardia corallina* (ATCC 19147). The yield of products varies depending on the amount of ion-exchange resin employed as indicated in Table 4. Most of the uses of ion-exchangers reported in the literature are for the removal of carboxylic acid products produced from hydrocarbon substrates. It is conceivable that

Table 4. Yield of toluic acids (*34* and *35*)[a] in the presence of ion exchange resins

Resin Used grams/liter	Yield of (*34*) and (*35*) grams/liter
0	0.1
21	3.8
45	4.5
90	5.5

[a] Formed from *p*-xylene incubated with *Nocardia corallina* (ATCC 19147) (see Refs. [78] and [79]).

H_3C–C₆H₄–CH_3 (33) ⟶ H_3C–C₆H₄–COOH (34) + H_3C–C₆H₂(OH)₂–COOH (35)

ion exchange resins could also be employed for the removal of a variety of other types of metabolites and starting materials in microbial transformation experiments. Many drugs yield phenolic metabolites which also have the potential to bind to ion exchangers. Examples include aporphines [80] and benzylisoquinolines like papaverine [15]. Other drugs and metabolites (e.g. those containing amine functions) should also be recoverable *via* ion-exchange resins.

When bacterial fermentations are used to perform biotransformations, resins may be recovered by simple filtration of the medium, or by low-speed centrifugation. Metabolites and unused starting materials may be reclaimed by extraction from the ion exchangers. However, when fungi, or actinomycetes are grown, the resulting mycelium may complicate the recovery process. The application of dialysis membrane as casings for ion-exchange resins offers the advantage of easy isolation of ion-exchange resins without having to remove large quantities of mycelium. Furthermore, resins would most likely be capable of being added to and removed from fermentation at time intervals to enable the reuse of cells in biotransformation reactions.

An alternative to the use of ion-exchange resins in the removal of substrates and microbial metabolites is the application of dialysis fermentation techniques. Apparatus design and the theory of dialysis fermentation have been reviewed [81]. Algae, bacteria, fungi, protozoa and tissue cells have all been studied using dialysis methods. Two or more compartments are typically employed in the dialysis culture apparatus; one is a culture compartment which is kept separate from the medium-containing reservoir compartment by a semipermeable membrane. Metabolites produced by fermentations diffuse from the culture chamber into the reservoir chamber. Efficient diffusion of metabolite to the reservoir is essentially dependent on the volume of reservoir fluid relative to the volume of the culture.

Dialysis devices have consisted of carboys containing dialysis membranes, special flasks containing reservoir and culture chambers, or fermentors containing dialysis tubing through which fluids are pumped from a distant reservoir and fermentor compartments holding fixed volumes of medium; the reservoir and fermentor can also be operated in a continuous fashion. Alternatively, batch reservoirs can be operated with continuous fermentors or *vice versa.* A dialysis fermentation unit with dialysis membrane isolated from both the fermentation tank and the reservoir is depicted in Fig. 3. In this apparatus, fluids are caused to pass through tubing to the dialysis chamber where exchange of nutrients or metabolites occur.

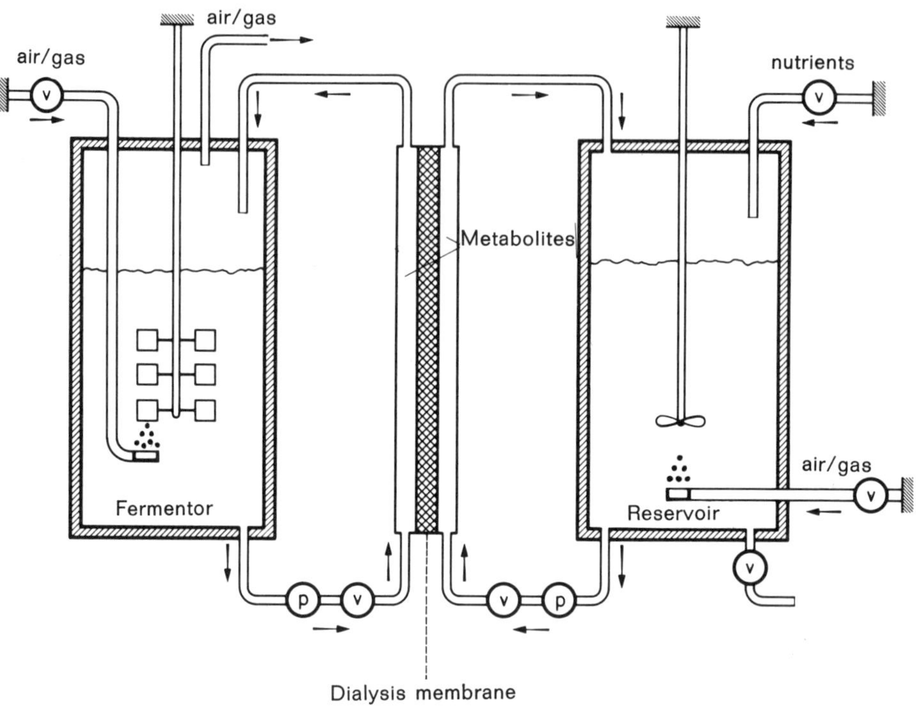

Fig. 3. Typical "dialyzer-dialysis system" described by Schultz and Gerhardt [81]. Permutations on flow regulation of nutrients and fermentation broths have been discussed in the literature ([81] and references cited therein). ⓟ pump; ⓥ valve

Several types of membranes have been used in dialysis culture. Ultramicroporous membranes with average pore sizes of 5 nm, permit passage of sugars and salts and some small microbial metabolites. Larger molecules such as proteins, polysaccharides, and microbial cells are excluded by such membranes. Microporous membranes with mean pore diameters of 200 nm down to about 25 nm retain species the size of bacteria, but permit most solutes including some macromolecules to pass through. Precautions must be taken to insure that flawless membranes are used to prevent the passage of bacteria from the fermentor to the reservoir. This is not as serious a problem with fungi since mycelia are considerably larger.

Vessels for use in differential dialysis separation of macromolecular products have been designed where two membranes are used to separate three chambers [82, 83]. One membrane allows permeation of macromolecules, but not of bacteria, while the second membrane allows small molecules to permeate, but not the macromolecules. Gerhardt and Gallup [84] described a dialysis flask which has been used in shaken-flask culture studies. The flask was useful in performing a number of operations including: producing dense populations of cells, preparing concentrated solutions of macromolecular products, removing toxic products which limit growth of biotransformations, and for supplying nutrients such as salts, carbohydrates and oxygen which would otherwise be depleted from growing cultures.

Dialysis fermentor systems have been used to concentrate cultures of microorganisms. Systems with large reservoirs and fermentors connected to one another are useful for this purpose but are not practical in the isolation of microbial metabolites due to the large size of the reservoir [86].

The application of dialysis fermentation to the removal of toxic fermentation end products has been reported. Cycloheximide produced by *Streptomyces griseus* inhibits its own production. It is also degraded in the absence of glucose [86]. *Streptomyces griseus* grown in a stirred fermentor was connected to a reservoir containing methylene chloride layered over with water by dialysis tubing which was coiled about the baffles in the fermentor. By pumping the aqueous dialysis fluid continuously through the fermentation, and then through a second chamber containing methylene chloride, cycloheximide was efficiently produced and extracted. After ten days of fermentation, production of cycloheximide doubled relative to the amounts produced under typical fermentation conditions. Furthermore, cycloheximide represented 82% of the solids in the methylene chloride in the extractor, and 60% of the total cycloheximide produced appeared in the organic reservoir fluid. From this example, one can conceive of a reservoir consisting of ion-exchange resins over which a dialysis fluid could be pumped. In this case, metabolites would be irreversibly bound to the ion-exchange resin, thereby obviating difficulties encountered with solvent in the reservoir.

Dialysis fermentation techniques have also been applied to the production of salicyclic acid from naphthalene by *Pseudomonas fluorescens* [87, 88]. Using a two-chamber dialysis culture flask, the amount of salicylate obtained from *P. fluorescens* (NRRL 3177) was abount twenty-fold greater than the quantity obtained in a control fermentation.

4. Cellular Models of Mammalian Metabolism

Mammals possess the ability to chemically alter a wide array of drugs and other xenobiotics. Most transformations occur in the liver though kidney, lung, and other tissues possess enzymes commonly involved in biotransformations. Because of the pre-eminent importance of hepatic tissue in xenobiotic metabolism, significant attempts have been made to define *in vitro* systems for more intensive study.

4.1 Comparison of Hepatic Systems

Many attempts have been made to define *in vitro* systems that could be used to study the metabolic fate of drugs and other chemicals in mammalian organisms. Much of this work has been performed with liver microsomal fractions and some success has been achieved in solubilizing enzyme systems that affect important transformations. However, qualitative differences between *in vitro* and *in vivo* systems continue to be observed. These apparent discrepancies could arise because of physicochemical changes of proteinaceous materials occurring during microsomal or enzyme isolation work. In addition, several workers have proposed that mitochondria (absent in most microsomal preparations) are involved in the control of hepatic cell mixed-function oxidations [89–91] through the transfer of reducing equivalents originating from NADPH or NADH through cytochrome b_5 to cytochrome P-450 during oxiation reactions catalyzed by hepatic microsomes. These workers showed that there was an increased rate of N-demethylation of ethylmorphine in the presence of mitochondria and that the rate of metabolism was slowed when mitochondrial participation was blockes. Furthermore, Schenkman *et al.* [91] reported that 50–70% of microsomal protein from the endoplasmic reticulum may be lost during standard fractionating procedures used to isolate microsomal fractions. Perfused liver systems, liver slices, and isolated liver cells in suspension have also been cited for their value as experimental models which span the gap between *in vivo* studies with animals and *in vitro* investigations with solubilized enzymes or microsomal preparations. However, when these three techniques are examined more closely, several disadvantages are uncovered.

The isolated perfused liver, although used extensively for biochemical studies [92], has several drawbacks when utilized as a model for drug metabolism. First, the viability of liver tissue is difficult to maintain for prolonged incubation periods while the apparatus necessary to perform perfusion experiments is complex and costly. Secondly, histological evaluation of the functional integrity of liver tissue is tedious and time-consuming. Thirdly, statistical problems may arise because of the small number of liver preparations that can be simultaneously treated in a given study.

Arguments against the use of liver slices as models for drug metabolism can be similarly presented. With liver slices, expensive cofactors are almost always necessary, leakage of pyridine and adenine nucleotides may occur, and depending on the thickness of the slices, problems with adequate oxygen diffusion and substrate penetration to all parts of the tissue may prevail with concommitant injury and destruction of cells [93, 94].

Isolated liver cells in suspension have recently been cited as valuable tools for drug metabolism studies [93–101]. However, as with the previously mentioned methods, there are potential problems associated with these preparations. A major area of concern with isolated cells is their membrane integrity and leakage of cofactors and intracellular enzymes. Hupka and Karlar [99] have reported a leakage of glucose 6-phosphate dehydrogenase from isolated cells into the incubation medium, while Junge and Brand [100] have noted a leakage of lactic dehydrogenase. Several investigators have suggested that isolated cells lose their ability to synthesize necessary cofactors for drug metabolism during incubation with foreign compounds [95] or that there is a loss of cofactors into the incubation media [93, 97]. Another problem with isolated cells is a decrease in viability with time after preparation of the cells. For example, Moldeus *et al.* [94] have observed that from two to five hours after isolation of the cells, viability decreased from 50 to 90%. Other workers have reported extreme variation in viability immediately after isolation, ranging from 50 to 60% [95]; 60 to 90% [100]; and 75 to 95% [101]. To increase the viability and metabolic activity of the cells, some investigators have resorted to the use of high concentrations of serum or albumin in the incubation medium. Inaba and coworkers [96] used as much as 17.5% v/v of horse serum in the incubation medium, while Gerayesh-Nejad *et al.* [97] added significant amounts of bovine serum albumin to the incubation assay mixture. Cultures of hepatic cells are normally grown in media containing 5 to 10% serum. A potential problem with increased amounts of serum proteins in the incubation medium is a greater degree of drug binding to these proteins and hence a decrease in the availability of the drug to the cells. For example, the uptake and metabolism of sulfobromophthalein by isolated hepatic cells were inversely related to the concentration of serum or albumin in the incubation medium [102].

Many of the above-mentioned shortcomings can potentially be obviated through the use of primary hepatic cell cultures. The latter permit interactions between xenobiotics and drug-metabolizing enzymes in intact cells under nearly normal physiological conditions. Furthermore, concurrent toxicological evaluations of foreign organic compounds should be possible through use of the primary cell culture.

The value of primary cell cultures for studying the metabolism of drugs and evaluating the toxicity of various chemicals has been demonstrated in scattered reports. Pomerat and Leake [103] first proposed the use of tissue cultures for the routine toxicity testing of new drugs. They demonstrated that antihistamines were highly toxic to skin cultures, confirming the clinical observation of skin injury from topical application. Gabliks and Friedman [104] determined the ID_{50} (defined by them as the dose inhibiting growth of 50% of the cells) for eleven poisons, employing cell lines derived from human and mouse tissues. Cell culture methodology has long been recognized as a screening technique in cancer chemotherapy [105]. Evaluating almost 2000 compounds, Schepartz *et al.* [106] determined that a correlation between cytotoxicity *in vitro* and antitumor activity in rats existed and that the percentage yield of compounds active *in vivo* increased significantly if an *in vitro* pre-screen was first carried out. On this basis, the Cancer Chemotherapy National Service Center has adopted the use of cell culture for the routine screening of new carcinostatic agents for cytotoxic activity. In other studies, North and Menzer [107a] examined the influence of organophosphorus insecticides on the ability of mouse L-29 fibroblast cells to hydrolyze choline and phenyl ester. Ames *et al.* [107b] have devised a test for mutagenicity and carcinogenicity based on frameshift mutations observed with *Salmonella* histidine mutants grown in the presence of certain organic compounds that are activated by preincubation with an hepatic microsomal preparation. Palmer *et al.* [108] studied, among other effects, the mutagenicity and cytogenetic effects of DDT and some of its derivatives on an established cell line taken from the rat kangaroo. Established cell lines are, of course, easier to maintain and work with than primary cultures but the latter systems permit the use of cells that are obviously closer in their biochemical and physiological responses to cells of the same origin in the intact animal. Findings obtained with primary cultures should be regarded as more reliable for predicting the anticipated effects under *in vivo* conditions.

4.2 Growth and Maintenance of Primary Hepatic Cell Cultures

The potential for utilization of primary cultures of liver cells for drug metabolism or toxicity studies has remained largely untapped. A principal reason for this lack of research activity has been the difficulty of obtaining relatively pure cultures of liver parenchymal cells or hepatocytes. While a majority of the cells found in the mammalian liver are parenchymal, as much as 35% of the cell population is non-parenchymal, primarily reticuloendothelial in nature (Kupffer cells) [109]. Therefore, contamination of cultured hepatocytes with reticuloendotheial cells is a common occurence. The difficulty of obtaining pure cultures of hepatocytes was highlighted as recently as 1972 in a workshop on cultured liver cells [110]. This meeting of established tissue culture investigators did not reach a clear-cut consensus on the proper isolation or identification of cultured hepatocytes. Only in the last few years have more precise procedures for isolating relatively pure cultures of hepatocytes been developed [111–113].

Three types of liver tissue have been utilized for the preparation of primary monolayer cultures of hepatocytes: fetal [111, 114], neonatal [115, 116], or adult rat liver [113, 117]. In our laboratories, we have selected neonatal tissue as the source of hepatocytes for primary monolayer cultures. A modification of the procedure of Leffert and Paul [111] is utilized to prepare cultures of differentiated neonatal rat liver cells. This technique permits the growth of functional, parenchymal hepatocytes, but suppresses the growth of nonparenchymal liver cells. The selection method is based upon the ability of parenchymal liver cells to survive in arginine-deficient medium, while nonarginine-synthesizing cells are selectively weeded out of the cultures.

Fetal liver was not chosen as a source of tissue to prepare primary cultures of hepatocytes because fetal liver cells rapidly grow and divide, which is not characteristic of young or adult liver cells. Furthermore, fetal liver cells lack some of the drug-metabolizing enzyme systems found in adult liver cells. To overcome these deficiencies, we are using neonatal rats as a source of hepatocytes [115]. Neonatal rats, three to eight days old, have sufficiently high levels of smooth endoplasmic reticulum and cytochrome P-450 [118, 119] compared to adult tissue. There is also a concomitant increase of NADH cytochrome reductase after birth.

Primary cultures of non-dividing, parenchymal, adult rat liver cells would seem to be the ideal *in vitro* model to study drug metabolism but these cultures have one important disadvantage in that the cells are viable for only 7 days [120, 121]. Thus, long-term studies with these cells are difficult to perform and interpretation of data is obscured with cultures that are in a state of metabolic decline. Other problems associated with the use of primary cultures of adult liver cells are preparative and technical difficulties of the Berry and Friend [122] procedure commonly utilized to isolate these cells. That is, the procedure involves a closed circuit perfusion of the liver which is technically difficult and subjects the liver to increased chances of contamination. Additionally, the equipment needed is costly and highly specialized. During *in situ* perfusion of the liver, it is exposed to high concentrations of enzymes for prolonged periods and to low temperature (4° C). This may explain the fact that plasma membranes obtained from isolated liver cells are dissimilar to those obtained from the whole liver [123].

The preparation of primary cultures of neonatal rat liver cells does not entail the use of complex perfusion equipment and is much simpler to carry out. Essentially, the cells are isolated by repeated trypsinization of the tissue, with the resulting cell suspension plated into tissue culture dishes. In contrast to most cultures of adult liver cells, we have cultured monolayers of neonatal liver cells which have maintained their viability and capacity to metabolize drugs for up to five weeks [115]. Therefore, our procedure allows for detailed, chronic examination of metabolic properties of hepatic cells in culture without the fear of cellular degeneration complicating the data. This longevity is an important feature when one considers the metabolism of drugs to metabolites which may be toxic. A comparison of the metabolism and toxicity of various agents may easily be evaluated in the same culture system.

4.3 Use of Primary Hepatic Cell Cultures in Xenobiotic Metabolism Studies

The few studies where cell culture systems have been used to investigate drug metabolism have involved mainly embryonic or fetal liver cells [124–127]. Of the studies with fetal liver cells, Gielen and Nebert [124, 125, 128] have conducted the most thorough investigations on mixed-function oxygenases. Specifically, these investigators have demonstrated that cultures of fetal hepatic cells can be induced by phenobarbital and polycyclic aromatic hydrocarbons to synthesize aryl hydrocarbon hydroxylase. Although much has been learned about hydroxylase induction in cultured cells, these studies have been limited to the metabolism of polycyclic hydrocarbons and the importance of the formation of intermediate epoxide metabolites in carcinogenesis. Other reports of metabolism studies with primary hepatic cultures include: biotransformation of porphyrin-inducing drugs in chick embryo liver cell cultures [129] and conjugation of bilirubin by cultures rat hepatoma cells [130]. A more definitive and basic study on *in vitro* biotransformation reactions was conducted by Poland and Kappas [127, 131] who demonstrated that aminopyrine and chlorcyclizine were demethylated in chick embryonic liver cultures in a manner similar to intact anmicals and by liver microsomal systems. Beyond these somewhat specialized investigations, no systematic studies on basic metabolic transformations, such as aromatic hydroxylation, O-dealkylation, N-dealkylation, N-oxidation, and S-oxidation, have been conducted in cultured hepatic cells, even though the potential for these systems to conduct various types of biotransformations has been suggested [127, 129–131].

It is significant that there have not been any major metabolism studies in primary cultures of young or adult hepatic cells. The only reference in the literature that describes the use of primary monolayer cultures of adult hepatocytes to study drug metabolism is an abstract by Pitot *et al.* [132]. One drawback associated with their technique of preparing and maintaining liver cell cultures is that exogenously-added insulin is necessary to preserve the viability and metabolic activity of the cells [117, 132]. Thus, the presence of insulin in the media could possibly interfere with the metabolism and activity of drugs being tested in the cultures. Furthermore, in order to obtain viable hepatic cells up to 20 days in culture, the author's technique required floating collagen membranes to which the cells attach and form a monolayer [117]. The preparation of these membranes adds another step to the already complex procedure of isolating the cells by the method of Berry and Friend [122].

In our laboratory, we have attempted to minimize as many procedural and technical problems as possible which may arise during the culturing of functional, parenchymal liver cells [115]. Our arguments for the use of neonatal rats as a source of liver tissue have been presented above. We believe that sufficient information is available in the literature to demonstrate the feasibility of developing rat hepatic cell cultures as mammalian models of drug metabolism. Our approach with hepatic cell cultures is to determine their ability to metabolize very simple compounds which serve as substrates for the classical metabolic reactions which occur in the liver: hydroxylation, N-dealkylation, O-dealkylation, N-oxidation, and S-oxidation. These reactions are of considerable importance in drug metabolism studies because of the involvement of similar functional groups in therapeutic agents. Most of these oxidative biotransformations are catalyzed by cytochrome P-450-linked monooxygenases, located primarily in the liver of mammals. Cytochrome P-450 monooxygenase are believed to affect aromatic hydroxylations, N- and O-dealkylations in mammals; S-dealkylations are probably mediated by similar enzymes [1]. Cultured hepatic cells [125, 128, 131, 132] possess monooxygenase enzyme systems similar to the cytochrome P-450 oxidizing systems of adult mammalian liver. Just as these oxidative reactions may be duplicated in microsomal fractions, it is very likely that they may be duplicated in cultured hepatocytes.

4.4 Hepatotoxicity Studies with Primary Liver Cell Cultures

Early liver cell injury produced by drugs or toxicants is extremely difficult to assess *in vitro*. The response of the liver to toxins or disease *in vivo* can be complicated by interactions between adjacent cells of different types, neural and hormonal feedback, and rapidly changing concentration of the agent at the cell-body fluid interface. With cultured cells as models of drug toxicity, neural and hormonal influences may be eliminated, and the toxicity of a given agent may evaluated directly on intact, viable cells. The study of cytotoxicity in cultured cells has previously relied on cell death as the end-point of damage produced by the offending agents. Other indicators of cell injury include changes in total protein, DNA or RNA content, morphological and ultrastructural alterations and changes in biochemical activity of selected enzymes. These indices of cell injury reflect obvious and gross changes in cell function and appearance. However, they do not readily explain the mechanism of injury produced by the offending agents. A more sensitive index of cell injury is an increase in the permeability of cellular membranes [133]. Because normal functioning of a cell depends upon the integrity of its organelles, cytochemical assessment of permeability changes in lysosomal and mitochondrial membranes can be used as a measure of early cell injury or cytotoxicity. Bitensky [134] first introduced the concept of lysosomal fragility as a means of measuring cell injury produced by toxic agents. Bitensky's technique, which was initially used as a histochemical evaluation of tissue sections, has been modified for use with cultured cells [135–138]. These studies have demonstrated that the toxic effects of many substances may be correlated with alterations in lysosomal membrane integrity.

Because mitochondrial structure and function are closely interrelated, cytochemistry has become increasingly employed to detect changes in mitochondrial integrity. Chayen and Bitensky [139] have suggested that "measurement of mitochondrial permeability or of the degree of activation of intramitochondrial enzymes would form a valuable guide to early cell damage." To detect toxicity or injury at the cellular level, we have developed sensitive cytochemical techniques which measure permeability changes of *in situ* lysosomes and mitochondria in cultured cells derived from the rat heart [140, 141]. We propose that these techniques may similarly be employed with primary hepatic cell cultures. Our techniques for measuring changes in mitochondrial and lysosomal membrane permeability are based on the principle of structure-linked enzyme latency [142, 143]. A lysosomal or mitochondrial enzyme is said to be latent if it is inactive towards its added substrates with intact organelles but not with disrupted or injured ones. Latency is most readily envisioned in terms of a permeability barrier lying between the sequestered enzyme and its substrates. Verity and Brown [143] have convincingly demonstrated that activation of latent enzyme activity in both mitochondria and lysosomes is a function of the membrane permeability of each organelle. Whereas intact, undamaged lysosomes and mitochondria allow limited quantities of substrates to reach their respective enzymes, injured organelles allow more of the substrates to penetrate and to react with the sequestered enzymes. Therefore, cytochemical staining of treated cells greater than that of controls is evidence of increased substrate accessibility to the latent enzymes and suggestive evidence of injury to the organelle membrane.

The prospect of subcellular organelles has usually been studied with either fixed or unfixed tissue sections, or with isolated organelles. Effective use of organelle permeability tests requires a minimum of damage to lysosomes and mitochondria during preparative procedures. Fixation, freezing, and sectioning [144] or homogenization and centrifugation procedures to obtain isolated organelles [133] may adversely affect the integrity of organelle membranes and thereby may reduce the reproducibility and sensitivity of the tests. To eliminate or minimize any procedural effects, as well as to maintain the *in situ* functional relationship of the organelles to the intact cell, we have chosen the cultured cell as the experimental model to study hepatic injury.

5. Outlook

5.1 Microbial Models of Mammalian Metabolism

The prospects for widespread use of microorganisms in xenobiotic metabolism studies appear very bright. Procurement of sufficient amounts of structurally complex drug metabolites for structure elucidation, and especially for biological testing has been a significant problem for a number of years. Reviews of the literature [1, 2] suggest that reliable microbial systems can be defined for the preparative synthesis of metabolites. In this review, we have tried to outline newer methodology and experimental approaches that can facilitate and improve the selectivity and yield of microbial transformations.

It is anticipated that considerable extensions of these efforts will be attempted in the future. The use of nonaqueous media, polymer-entrapped cells, and ion-exchange and dialysis techniques to remove inhibitory products appear to represent particularly fruitful approaches and should be further explored.
As comparison studies between microbial and mammalian transformations are pursued further, it seems highly probable that close parallels will be uncovered. After the microbial systems have been more fully defined, it is conceivable that they could be employed to predict metabolic patterns of new chemical entities in mammals. That is, microbially produced metabolites could serve as reference metabolites in subsequent mammalian investigations.

5.2 Cellular Models of Mammalian Metabolism

The primary hepatic cell culture systems described in this review are in a relatively early stage of development. However, it appears that systems are almost at hand that will metabolize drugs and other foreign chemicals in parallel fashion to that of intact animals. Since primary hepatic cell cultures can be maintained for long periods of time, they possess the potential of becoming exceptional tools for the study of hepatotoxicity. A number of drugs such as papaverine [145], acetaminophen [146] and furosemide [147] have poorly understood liver toxicities associated with their use either at high doses or for extended periods of time. Studies of agents like these and the underlying mechanisms of hepatotoxicity may be uncovered through the use of primary hepatic cell cultures.

Acknowledgements

This work was supported in part by Grant NS-12259, National Institute of Neurological and Communicative Disorders and Stroke, Grant CA-13786, National Cancer Institute and a Grant from the Pharmaceutical Manufacturers Association.

References

1. Smith, R. V., Rosazza, J. P.: J. Pharm. Sci **64**, 1737 (1975).
2. Smith, R. V., Rosazza, J. P.: Biotechnol. Bioeng. **17**, 785 (1975).
3. Peterson, D. H., Murrah, H. C.: J. Amer. Chem. Soc. **74**, 1871 (1952).
4. Iizuka, H., Naito, A.: Microbial Transformation of Steroids and Alkaloids. State College, Penn.: University Park Press 1967.
5. Charney, W., Herzog, H. L.: Microbial Transformations of Steroids. New York: Academic Press 1967.

6. Beukers, R., Marx, A. F., Zuidweg, M. H. J.: In: Drug Design, Vol. 3, E. J. Ariens, Ed., pp. 1–117. New York: Academic Press 1972.
7. a) Fonken, G., Johnson, R. S.: Chemical Oxidations with Microorganisms. Marcel Dekker. New York: 1972;
   b) Skryabin, G. K., Golovleva, L. A. M.: Microorganisms in Organic Chemistry. Moscow: Nauka 1976;
   c) Kieslich, K.: Microbial Transformations of Non-Steroid Cyclic Compounds. New York: Wiley 1976.
8. Ferris, J. P., Fasco, M. J., Stylianopoulou, F. L., Jerina, D. M., Daly, J. W., Jeffrey, A. M.: Arch. Biochem. Biophys. **156**, 97 (1973).
9. Gunsalus, I. C., Meeks, J. R., Lipscomb, J. D., DeBrunner, P., Munck, E.: In: Molecular Mechanisms of Oxygen Activation (O. Hayaishi, Ed.), pp. 559–613. New York: Academic Press 1974.
10. Duppel, W., LeBeault, J. M., Coon, M. J.: Euro. J. Biochem. **36**, 583 (1973).
11. Gallo, M., Roche, B., Azoulay, E.: Biochim. Biophys. Acta **419**, 425 (1976).
12. Ambike, S. H., Baxter, R. M.: Phytochemistry **9**, 1959 (1970).
13. Weaver, E. A., Kenney, H. E., Wall, M. E.: Appl. Microbiol. **8**, 345 (1960).
14. Betts, R. E., Walters, D. E., Rosazza, J. P.: J. Med. Chem. **17**, 599 (1974).
15. Rosazza, J. P., Kammer, M., Youell, L., Smith, R. V., Erhardt, P. W., Truong, D. H., Leslie, S. W.: Xenobiotica (Submitted for Publication).
16. Davis, P. J., Rosazza, J. P.: J. Org. Chem. (In Press, July 1976).
17. Jones, I. B.: Applications of Biochemical Systems in Preparative Organic Chemistry. New York: Wiley 1976.
18. Wallen, L. L., Stodola, F. H., Jackson, R. W.: Type Reactions in Fermentation Chemistry, Bull. ARS-71-13, Washington, D. C., Agricultural Research Service U.S. Department of Agriculture: 1959.
19. Marsheck, W. J., Jr.: Progress Industr. Microbiol. **10**, 49 (1971).
20. Karrow, E. O., Petsiavas, D. W.: Indust. Engin. Chem. **48**, 2213 (1956).
21. Martin, S. M., Skerman, V. B. D.: World Directory of Collections of Cultures of Microorganisms. New York: Wiley 1972.
22. Marshall, V., Wiley, P. F.: J. Antibiot. **28**, 838 (1975).
23. Munnecke, D. M., Hsieh, D. P. H.: Appl. Microbiol. **30**, 575 (1975).
24. Tay, L. K., Sinsheimer, J. E.: J. Pharm. Sci. **64**, 471 (1975).
25. Yang, R. D., Humphrey, A. E.: Biotechnol. Bioeng. **17**, 1211 (1975).
26. Caldwell, J. M., Hawksworth, G. M.: J. Pharm. Pharmacol. **25**, 422 (1973).
27. Lee, B. K., Brown, W. E., Ryu, D. Y., Jacobson, H., Thoma, R. W.: J. Gen. Microbiol. **61**, 97 (1970).
28. Murray, H. C., Peterson, D. H.: U.S. Patent 2,602,769 (1952); Chem. Abstr. **46**, 8331 f.
29. Dulaney, E. L., McAleer, W.: U.S. Patent 2,802,775 (1957); Chem. Abstr. **52**, 1299e; Rolls, J. P. and Knight, J. C., Abstracts, Amer. Chem. Soc. Meeting **164**, MICR 14 (1972).
30. a) Lefebvre, G., Schneider, F., Germain, P., Gay, R.: Tetrahedron Lett., 127 (1974);
    b) ATCC Culture Catalogue, Catalogue of Strains, 12th ed., Rockville, MD., American Type Culture Collection: 1976.
31. Smith, R. V., Rosazza, J. P., Nelson, R. A.: J. Chromatogr. **95**, 247 (1974).
32. Smith, R. V., Rosazza, J. P., Engel, K. O., Humphrey, D. W.: J. Chromatogr. **106**, 235 (1975).
33. Kalyanaraman, V. S.: J. Scien. Indus. Res. **30**, 560 (1971).
34. Demain, A. L.: Advan. Biochem. Eng. **1**, 113 (1970).
35. Wiseman, A., Lim, T.-K., McCloud, C.: Biochem. Soc. Trans. **3**, 276 (1975).
36. Horvath, R. S.: Bacteriol. Rev. **36**, 146 (1972).
37. Wodzinski, R. S., Coyle, F. E.: Appl. Microbiol. **27**, 1081 (1974).
38. Klevens, H. B.: J. Phys. Chem. **54**, 283 (1950).
39. Wodzinski, R. S., Bertolini, D.: Appl. Microbiol. **23**, 1077 (1972).
40. Jones, J. B., Baskevitch, N.: Steroids **22**, 525 (1973).

41. Kondo, E., Masuo, E.: J. Gen. Appl. Microbiol. **7**, 113 (1961).
42. Raymond, R. L.: U.S. Patent 3,244,946 (1965); Chem. Abstr. **64**, 7327c.
43. Weaver, E. A.: British Patent 912,274 (1962); Chem. Abstr. **58**, 13093e.
44. Cooney, C. L., Hueter, J.: Biotechnol. Bioeng. **16**, 1045 (1974).
45. Barth, T., Jost, K., Rychlik, I.: Coll. Czech. Chem. Commun. **38**, 2011 (1973).
46. Cremonesi, P., Carrea, G., Ferrara, L., Antonini, E.: Eur. J. Biochem. **44**, 401 (1974).
47. Cremonesi, P., Carrea, G., Ferrara, L., Antonini, E.: Biotechnol. Bioeng. **12**, 1101 (1975).
48. Cremonesi, P., Carrea, G., Sportoletti, G., Antonini, E.: Arch. Biochem. Biophys. **159**, 7 (1973).
49. Lugaro, G., Carrea, G., Cremonesi, P., Casellato, M., Antonini, E.: Arch. Biochem. Biophys. **159**, 1 (1973).
50. Buckland, B. C., Dunnill, P., Lilly, M. D.: Biotechnol. Bioeng. **17**, 815 (1975).
51. McGregor, W. C., Tabenkin, B., Jenkins, E., Epps, R.: Biotechnol. Bioeng. **14**, 831 (1972).
52. Valenta, J. R., DiCuollo, C. J., Fare, L. R., Miller, J. A., Pagano, J. F.: Appl. Microbiol. **28**, 995 (1974).
53. Heusghem, C., Welsch, M.: Bull. Soc. Chim. Biol. **31**, 282 (1949).
54. Talalay, P., Dobson, M. M., Tapley, D. F.: Nature **170**, 620 (1952).
55. Dulaney, E. L., Stapley, E. O., Hlavac, C.: Mycologia **47**, 464 (1955).
56. Shull, G. M., Kita, D. A.: J. Amer. Chem. Soc. **77**, 763 (1955).
57. Vezina, C., Seghal, S. N., Singh, K.: Appl. Microbiol. **11**, 50 (1962).
58. Lee, B. K., Brown, W. E., Ryu, D. Y., Thoma, R. W.: Biotechnol. Bioeng. **13**, 503 (1971).
59. Davis, P. J., Gustafson, M. E., Rosazza, J. P.: Biochim. Biophys. Acta **385**, 133 (1975).
60. Wix, G., Albrecht, K.: Nature **183**, 1279 (1959).
61. Silman, I. H., Katchalski, E.: Ann. Rev. Biochem. **35**, 873 (1966).
62. Kolarik, M. J., Chen, B. J., Emery, A. H., Jr., Lim, H. C.: In: Immobilized Enzymes in Food and Microbial Processes (A. C. Olson and C. L. Cooney, Eds.), p. 71. New York: Plenum Press 1974.
63. Skinner, K. J.: Chem. Eng. News **53** (33), 22 (1975).
64. Sticks, G. P., Updike, S. J.: Anal. Chem. **38**, 726 (1966).
65. Wieland, T., Determan, H., Buening, K.: Z. Naturforschung **21**b, 1003 (1966).
66. Leuschner, F.: British Patent 953,414 (1964); Chem. Abstr. **62**, 760c.
67. Mosbach, K., Mosbach, R.: Acta Chem. Scand. **20**, 2807 (1966).
68. Mosbach, K., Larsson, P-O.: Biotechnol. Bioeng. **12**, 19 (1970).
69. Vieth, W. R., Wang, S. S., Saini, R.: Biotechnol. Bioeng. **15**, 565 (1973).
70. Yamamoto, K., Sato, T., Tosa, T., Chibata, I.: Biotechnol. Bioeng. **16**, 1589 (1974).
71. Franks, N. E.: Biotechnol. Bioeng. Symp. **3**, 327 (1972).
72. Tosa, T., Sato, T., Mori, T., Chibata, I.: Appl. Microbiol. **27**, 886 (1974).
73. Chibata, I., Tosa, T., Sato, T.: Appl. Microbiol. **27**, 878 (1974).
74. Kosinkiewicz, B.: J. Chromatogr. **114**, 463 (1975).
75. Schwartz, J., Margalith, P.: Biotechnol. Bioeng. **15**, 85 (1973).
76. Tone, H., Kitai, A., Ozaki, A.: Biotechnol. Bioeng. **10**, 689 (1968).
77. Kitai, A., Tone, H., Ishikura, I., Ozaki, A.: J. Ferm. Technol. **46**, 442 (1968).
78. Humphrey, A. E., Raymond, R. L.: British Patent 1,111,310 (1968): Chem. Abstr. **69**, 17133t.
79. Raymond, R. L., Jamison, V. M., Hudson, J. O.: Appl. Microbiol. **17**, 512 (1969).
80. Rosazza, J. P., Smith, R. V., Stocklinski, A. W., Gustafson, M. E., Adrian, J.: J. Med. Chem. **18**, 791 (1975).
81. Schultz, J. S., Gerhardt, P.: Bacteriol. Rev. **33**, 1 (1969).
82. Herald, J. D., Schultz, J. S., Gerhardt, P.: Appl. Microbiol. **15**, 1192 (1967).
83. Gerhardt, P., Gallup, D. M.: U.S. Patent 3,186,917 (1962).
84. Gerhardt, P., Gallup, D. M.: J. Bacteriol. **86**, 919 (1963).
85. Gallup, D. M., Gerhardt, P.: Appl. Microbiol. **11**, 506 (1963).
86. Kominek, L. A.: Antimicrob. Agents Chemother. **7**, 861 (1975).
87. Abbott, B. J., Gerhardt, P.: Biotechnol. Bioeng. **12**, 577 (1970).
88. Abbott, G. J., Gerhardt, P.: Biotechnol. Bioeng. **12**, 591 (1970).

89. Hildebrandt, A., Estabrook, R. W.: Arch. Biochem. Biophys. **143**, 66 (1971).
90. Cinti, D. L., Schenkman, J. B.: Mol. Pharmacol. **8**, 327 (1972).
91. Schenkman, J. B., Cinti, D. L., Moldeus, P.: Ann. N. Y. Acad. Sci. **212**, 420 (1973).
92. Miller, L. L.: In: Isolated Liver Perfusion and Its Applications (I. Bartosik, A. Guaitani, L. L. Miller, Eds.) pp. 11–52. New York: Raven Press 1973.
93. Henderson, P. Th., Dewaide, J. H.: Biochem. Pharmacol. **18**, 2087 (1969).
94. Moldeus, P. Grundin, R., Vadi, H., Orrenius, S.: Eur. J. Biochem. **46**, 351 (1974).
95. Holtzman, J. L., Rothman, V., Margolis, S.: Biochem. Pharmacol. **21**, 581 (1972).
96. Inaba, T., Umeda, T., Mahon, W. A.: Life Sci. **16**, 1227 (1975).
97. Gerayesh-Nejad, S., Jones, R. S., Parke, D. V.: Biochem. Soc. Trans. **3**, 403 (1975).
98. Orrenius, S., Moldeus, P., Vadi, H., Grundin, R.: Biochem. Soc. Trans. **3**, 817 (1975).
99. Hupka, A. L., Karlar, R.: J. Reticuloendothel. Soc. **14**, 225 (1973).
100. Junge, O., Brand, K.: Arch. Biochem. Biophys. **171**, 398 (1975).
101. Howard, R. B., Pesch, L. A.: J. Biol. Chem. **243**, 3105 (1968).
102. Stege, T. E., Loose, L. D., DiLuzio, N. R.: Proc. Soc. Exp. Biol. Med. **149**, 455 (1975).
103. Pomerat, C. M., Leake, C. D.: Ann. N. Y. Acad. Sci. **58**, 1110 (1954).
104. Gabliks, J., Friedman, L.: Ann. N. Y. Acad. Sci. **160**, 254 (1969).
105. Ormsbee, R. A., Cornman, I.: Cancer Res. **8**, 384 (1948).
106. Schepartz, S., Macdonald, M., Leiter, J.: Proc. Amer. Assoc. Cancer Res. **3**, 265 (1961).
107. a) North, H. H., Menzer, R. E.: J. Agr. Food Chem. **18**, 797 (1970);
b) Ames, B. N., Durston, W. E., Yamasaki, E., Lee, F. D.: Proc. Nat. Acad. Sci., U.S.A. **70**, 2281 (1973).
108. Palmer, K. A., Green, S., Legator, M. S.: Toxicol. Appl. Pharmacol. **22**, 355 (1972).
109. Weibel, E. R., Staubil, W., Gnagi, H. R., Hess, F. A.: J. Cell Biol. **42**, 68 (1969).
110. Potter, V. R.: Cancer Res. **32**, 1998 (1972).
111. Leffert, H. L., Paul, D.: J. Cell Biol. **52**, 559 (1972).
112. Berry, M. N.: In: Methods in Enzymology, Vol. 32, Part B (Fleischer, S., Packer, L., Eds.), p. 625. New York: Academic Press 1974.
113. Bonney, R. J.: In Vitro **10**, 130 (1974).
114. Gerschenson, L. E., Berliner, J., Davidson, M. B.: In: Methods in Enzymology, Vol. 32, Part B (Fleischer, S., Packer, L., Eds.) p. 733. New York: Academic Press 1974.
115. Acosta, D., Smith, R. V., Anuforo, D., Baaske, D. M.: In Vitro (Submitted for Publication).
116. Armato, U., Draghi, E., Andreis, P. G.: Cell Differ. **4**, 147 (1975).
117. Michalopoulos, G., Pilot, H. C.: Exp. Cell Res. **94**, 70 (1975).
118. Dallner, G., Siekwitz, P., Palade, G. E.: J. Cell Biol. **30**, 73 (1966).
119. Dallner, G., Siekwitz, P., Palade, G. E.: J. Cell Biol. **30**, 97 (1966).
120. Bissell, D. M., Hammaker, L. E., Meyer, U. A.: J. Cell Biol. **59**, 722 (1973).
121. Bonney, R. J., Becker, J. E., Walker, P. R., Potter, V. R.: In Vitro **9**, 399 (1974).
122. Berry, M. N., Friend, D. S.: J. Cell Biol. **43**, 506 (1969).
123. Solyom, A., Lauter, C. J., Trams, E. G.: Biochim. Biophys. Acta **274**, 631 (1972).
124. Gielen, J. E., Nebert, D. W.: Science **172**, 167 (1971).
125. Gielen, J. E., Nebert, D. W.: J. Biol. Chem. **246**, 5189 (1971).
126. Pelkonen, O., Korhonen, P., Jouppila, P., Karki, N.: Life Sci. **16**, 1403 (1975).
127. Poland, A., Kappas, A.: Biochem. Pharmacol. **22**, 749 (1973).
128. Nebert, D. W., Gielen, J. E.: J. Biol. Chem. **246**, 5199 (1971).
129. Racz, W. J., Moffat, J. A.: Biochem. Pharmacol. **23**, 215 (1974).
130. Wolf, C. F. W., Munkelt, B. E., Kaighn, M. E.: Proc. Soc. Exp. Biol. Med. **145**, 918 (1974).
131. Poland, A., Kappas, A.: Molec. Pharmacol. **7**, 697 (1971).
132. Pitot, H. C., Michalopoulos, G., Sattler, C., Sattler, G.: Amer. J. Pathol. **82**, 47a (1976).
133. Chayen, J., Bitensky, L.: In: Cell Biology in Medicine (Bittar, E. E., Ed.) p. 595. New York: Wiley 1973.
134. Bitensky, L.: In: Lysosomes. Ciba Foundation Symposium (de Reuck, A. V. S., Cameron, M. P., Eds.), p. 362. London: Churchill 1963.
135. Maggi, V., Riddle, P. N.: J. Histochem. Cytochem. **13**, 310 (1965).

136. Fand I: Arch. Int., Pharmacodyn. **183**, 247 (1970).
137. Wenzel, D. G., Acosta, D.: Res. Commun. Chem. Pathol. Pharmacol. **6**, 689 (1973).
138. Reed, B. L., Wenzel, D. G.: Histochem. J. **7**, 115 (1975).
139. Chayen, J., Bitensky, L.: In: The Biological Basis of Medicine, Vol. 1 (Bittar, E. E., Bittar, N., Eds.), p. 337. New York: Academic Press 1968.
140. Acosta, D., Wenzel, D. G.: Atherosclerosis **20**, 417 (1974).
141. Acosta, D., Wenzel, D. G.: Histochem. J. **7**, 45 (1975).
142. Bendall, D. S., DeDuve, C.: Biochem. J. **74**, 444 (1960).
143. Verity, M. A., Brown, W. J.: Exp. Mol. Pathol **19**, 1 (1973).
144. Conklin, J. L., Dewey, M. M., Kahn, R. H.: Amer. J. Anat. **110**, 19 (1962).
145. Kiaer, H. W., Olson, S., Ronnov-Jessen, V.: Arch. Pathol. **98**, 292 (1974).
146. Mitchell, J. R., Jollow, D. J., Potter, W. Z., Davis, D. C., Gillette, J. R., Brodie, B. B.: J. Pharmacol. Exp. Therap. **187**, 185 (1973).
147. Mitchell, J. R., Jollow, D. J., Gillette, J. R., Brodie, B. B.: Drug Metab. Disp. **1**, 418 (1973).

The Characterization of Mixing in Fermenters

J. BRYANT
Dept. of Chemical Engineering, The University of Exeter, Exeter, U.K.

With 18 Figures

Contents

1. Introduction

The object of this contribution is to discuss mixing and agitation in relation to the performance of fermenters and its characterization by determination of the distribution of circulation rates.

One of the problems facing the designer of a fermenter is the choice of the correct form of mechanical agitation system. The problem is confused to the extent that what is required of the agitator is not always obvious, in the sense that the critical function of agitation for a particular fermentation is not generally known. Further confusion arises because performance characteristics of impellers, and the variation of these characteristics with the rheological properties of broth, are not clearly understood.

It is convenient to consider systems in which mechanical agitation is provided by a rotating stirrer, or stirrers, mounted on a shaft on the vertical centre-line of the fermenter, as is shown in Fig. 1. This is not a wholly restrictive assumption and much of what follows is equally applicable to other agitation systems such as air-lift fermenters, for example. It is, however, much easier to take one particular form of system as a basis for discussion.

In any fermentation process, agitation, often but not always coupled with aeration, has to perform and provide a variety of different functions. The skill of the designer is that he should be able to define unequivocally which of these functions is critical, or limit-

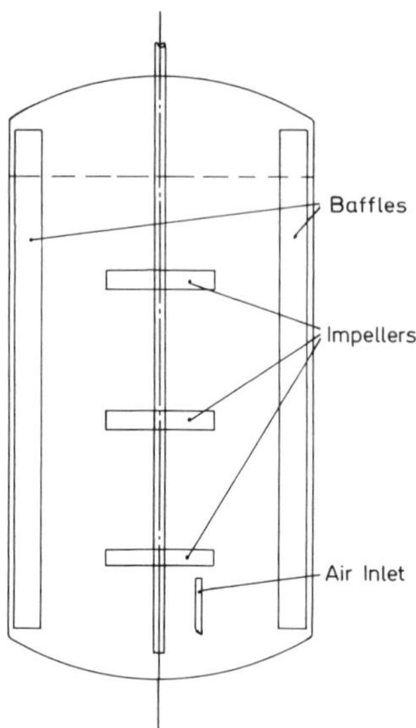

Fig. 1. Layout of typical stirred fermenter

ing performance of the fermentation, and to design accordingly. As an example of the analysis of the different functions of agitation the operation of a deep culture mycelial fermentation is considered.

One function of agitation is to suspend the mycelium uniformly throughout the working volume of the fermenter. Since the density of mycelium is little different from that of the supporting medium, suspension can be achieved with relative ease. Similarly trace elements and minor nutrients must also be uniformly distributed: provided the rates of consumption of these species are low, and this is an important proviso, this function of agitation can be satisfied at relatively low power inputs. A more demanding duty of the agitation system is to reduce temperature gradients in the bulk of the broth to acceptably small values, and to enhance heat transfer rates between the broth and cooling surfaces. Although this is more difficult to achieve than the corresponding reduction in concentration gradients it is still possible at low power inputs.

It is only when the provision of the main nutrients, sugar and oxygen, is considered that the critical function of agitation becomes apparent. There is a clear distinction between sugar, which is readily soluble in the broth, and oxygen with a solubility of ca. 10 p.p.m. under normal operating conditions. This difference is essentially one of time scale. The broth has, or can have, a large capacity for sugar, and at any time there can be enough in solution to satisfy the demands of the microorganisms for say 5–10 min or even for much longer periods. The capacity of the broth for dissolved oxygen is at least one order of magnitude less; with an active broth of the usual mycelial concentration its capacity may be only enough for 15–30 sec of normal respiration.

In this example then the supply of oxygen is clearly the most difficult function demanded of the agitation system. Provided this function is satisfied all other duties of the stirrer will be met. It is worth pursuing this discussion a little further, and to consider the supply of oxygen in rather more detail. The supply route comprises three stages; the solution of oxygen from the sparged air; the distribution of this oxygen throughout the broth and finally its transfer from the liquid to the mycelium. The problem is to determine which of these three stages is limiting or controlling the performance. There is no reason to suppose that any one particular stage, say gas-to-liquid transfer, is limiting everywhere in the fermenter. The more probable situation is that the stage controlling oxygen availability depends on position in the fermenter and furthermore that this position itself depends on the state of the fermentation. Around the impeller immediately above the air sparger the conditions of agitation are such that oxygen is freely available and the only limitation on performance is the strain of microorganisms being used. In a multiple-impeller system, such as shown in Fig. 1, it is probable that around the other impellers liquid-phase transport and liquid-to-solid transfer are satisfactory, but oxygen solution may not necessarily be sufficiently fast. Similar arguments can be used for other regions in the fermenter, such as at the edges of baffles or near the free surface of the broth.

This point has been emphasised to stress the important judgment needed at this stage in fermenter design. Should the object be to design the fermenter so that is behaves as a backmix reactor in which every piece of mycelium experiences an identical environment at any time? Alternatively, should possible spatial variations be recognised and accepted in the design? Perfect mixing can probably be approached closely in laboratory or bench scale fermenters, particularly those operating with bacterial cultures. In these circumstances it is quite proper to measure oxygen supply or availability in terms of a gas to liquid mass transfer coefficient, a dissolved oxygen concentration in the broth and a liquid to solid mass transfer coefficient. Those three measures each have unique values for a particular system, independent of position in the fermenter. The only doubt is whether the dissolved oxygen concentration as measured by a probe, typically a galvanic cell, is the same as that experienced by the microorganisms. The reason for this doubt is that if the dissolved oxygen probe is to function satisfactorily then it must be placed in a flowing medium and its boundary layer characteristics will be quite different from those of a freely suspended piece of mycelium or bacterial cell. This point is perhaps rather pedantic if the fermenter is perfectly mixed, but in a large fermenter it is a valid worry if mixing is not perfect and oxygen gradients are present, as may well be the case.

If a fermenter is not perfectly mixed then considerable problems arise when attempts are made to characterize mixing or agitation by means of oxygen availability. The overall supply rate can obviously be found from a balance across the inlet and outlet gas flows, but any meaningful measure of gas-to-liquid mass transfer coefficient is extraordinarily difficult to obtain. The reason for this is that if oxygen gradients exist then local transfer coefficients only have significance and any application if the flow pattern in the fermenter is known completely. To state the obvious, it is the environment experienced by the microorganisms that determines performance, and not the partial view of the same system provided by overall gas-to-liquid mass transfer coefficients or local measures of dissolved oxygen concentration.

In terms of characterizing the agitation conditions in a mycelial fermentation then a gas-to-liquid mass transfer coefficient is satisfactory if the fermenter is perfectly mixed and there is no evidence of liquid-to-solid transfer rates controlling. As soon as the scale increases so that it is no longer economically practicable to ensure perfect mixing alternative methods of characterization are required.

A more realistic approach to design is to accept that at all reasonable power inputs spatial variations are bound to exist within the broth. The problem of design then is to identify a region, or regions, within the fermenter where oxygen availability is more than enough to satisfy the demands of the microorganisms, and then to ensure that broth circulates through these regions at an acceptable frequency.

This brief discussion has centred on systems such as the penicillin fermentation, a highly aerobic process in a thick mycelial suspension, but it is equally valid for many other fermentation processes. It has shown the importance of identifying the critical function of agitation, and how this function may depend on position in the fermenter. Unless this critical stage is recognised and is involved in the method used to characterize agitation then such characterizations will be ill-founded and of little real value.

2. Methods of Characterizing Mixing

Before discussing any methods of characterising mixing or agitation it is worth considering some of the prerequisites of any method that is to have wide applicability. These prerequisites are:

1. the method and equipment should be capable of use with any fermenter from say 500 l capacity upwards;
2. the method should be applicable to a real fermentation;
3. the method should be continuous, or at least allow data to be obtained at regular intervals throughout a fermentation;
4. the results should be in a form that can be related simply to both fermentation performance and to fermenter design and operation.

Considering these prerequisites in more detail, it is obviously an advantage if the method is applicable to large fermenters as well as to pilot plant vessels. This ensures a simple interpretation of results from different scales of operation because the only variable has been agitation itself, there being no additional problems associated with change of method or technique.

If the technique is to be applied to a real fermentation it is absolutely essential that its use should introduce no risk of infection whatsoever, either during fermenter sterilisation or during the course of the fermentation. A further implication of this prerequisite is that the calibration of the characterizing equipment must be stable as it is unlikely to be possible to withdraw it from the fermenter during a run for recalibration. Finally, of course, the equipment must be robust to withstand the environment inside a well-agitated fermenter.

In many fermentations the rheological properties of the broth, and its volume, change with time: agitation is frequently provided by a fixed speed motor. In these circumstances the agitation conditions must change with the age of the fermentation and so some continuous assessment of these conditions should be available of the characterization is to be well-founded.

The final prerequisite is the most obvious and needs no justification or amplification. The assumption made in formulating these four prerequisites was that the fermentation was well-developed and being operated on a production scale in large fermenters, with the possibility of continuing investigational work in pilot plant fermenters. In these circumstances a well-founded quantitative characterization of mixing is essential if the full potential of the process is to be realised. However, many fermentations, particularly those in their early stages of development, are carried out in fermenters of say 5–20 l capacity. The problem of providing adequate mixing, and characterizing it, on this scale of operation is quite different. The prerequisites may be relaxed, not because they are irrelevant, but rather because there are probably still many aspects of the process to be developed quite apart from mixing.

At this stage of development there is a wide variety of methods available to characterize mixing, all of which provide useful preliminary design information, although they may not meet the prerequisites listed above.

Whatever the scale of operation, visual observation of the system is essential: the points to note are the nature of the flow, its pattern, the presence of any dead regions, and, if the system is aerated, the size and distribution of the bubbles.

The methods of characterization may be broadly classified as

a) stimulus/response techniques,
b) flow-followers and local velocity measurement,
c) miscellaneous.

One of the simplest applications of a stimulus/response method is to measure the terminal mixing time. At some time, taken as zero, a pulse of tracer is added to the system. The time at which the tracer is distributed to some arbitrary degree of uniformity is then taken as the terminal mixing time. The size of the pulse and the accepted degree of uniformity affect the actual value of this time. However, the method is simple and provides useful data.

If this method, or indeed any stimulus/response method, is to be used in a real fermentation broth there is always the problem of finding a suitable tracer. The tracer must not affect the course of the fermentation and not interfere with any subsequent extraction processes, and ideally its concentration should be capable of detection and measurement in an aseptic manner. The reason for wanting mixing data from a real broth is that in practice it is difficult to simulate all the rheological properties in an artificial medium. Not only must the properties of the unaerated broth be matched, but also the modifications to these properties when air is sparged into the system. This latter simulation is by far the more difficult to achieve.

Another problem associated with many stimulus/response methods is that successive tests on the same sample of liquid increase the background concentration of tracer and so reduce the sensitivity of the measurement.

One attractive method has been described (Blakebrough and Sambamurthy, 1966; McManamey *et al.*, 1973) that obviates some of the difficulties mentioned above. The basis of the method is to adjust the pH of the liquid to some known value in the alkaline range and add phenophthalein. The stimulus, or pulse, consists of a carefully measured equivalent of acid such that it is in slight excess. The time of disappearance of the last whisp of pink is then taken as the mixing time. The method gives reproducible results and can be used with aerated systems; repeated tests are possible without loss of sensitivity.

An extension of this method is to record the manner in which the concentration of tracer reaches uniformity, rather than only the time at which uniformity is achieved. Although the necessary apparatus is more complex, the results provide extra information. It is possible to measure the mean circulation time and the effective dispersion in the system in terms of a Peclet number (e.g. Holmes *et al.*, 1964).

Circulation time can be measured more conveniently by using inert flow-followers, typically small chips of plastics (e.g. Sykes, 1965).This method is limited to clear unaerated liquids. The mean circulation time is a direct measure of the pumping capacity of the impeller system and has been well-documented (Uhl and Gray, 1966). Followers incorporating radio isotopes have also been described (Aiba, 1958).

Local velocity measurements and photographic analysis of flow conditions and bubble formations have been described in the literature (Metzner and Taylor, 1960; van't Riet, 1975), but it is not fully evident how these various data can be related to design and performance.

Specific power input, that is the power input per unit volume of fluid, has been widely used. It is a convenient parameter for scale-up purposes, and in a later section of this contribution, its basic importance is discussed in more detail.

The general problems of mixing and its characterization have been well summarised (Aiba *et al.*, 1973; Finn, 1967) and they should be consulted for broad guidance.

A method of characterization involving the use of a flow-follower incorporating a radio transmitter is described in detail in the Appendix. Its advantage is that it meets all the prerequisites set out above, and the results obtained from its use can be related directly to design and performance. The interpretation of these results is considered in detail in the next section.

3. Interpretation of Results

An analysis of the frequency pattern for passage of a flow follower through an active zone will provide the following data:

a) Mean circulation time, $\bar{t}$;

b) Standard deviation of circulation times, σ;

c) Distribution of circulation times, $f(t)$.

$f(t)dt$ is the proportion of circulations whose times of circulation lie between t and $(t + dt)$.

These data may be related to fermentation performance and to fermenter design and operation.

3.1 Fermentation Performance

The first step in the treatment of the data is to plot $f(t)$ against t. Figure 2 shows a typical curve: in general it takes the form of a log-normal distribution. The next step is to calculate a characteristic reaction time T. If the limiting nutrient is consumed by a zero order reaction then

$$T = c_0/k_0 \tag{1}$$

and so T represents the time required to reduce the nutrient concentration from c_0 to zero in the absence of any further supply. For all other reaction orders the limiting nutrient concentration can only approach zero asymptotically. For instance if nutrient is consumed by a reaction with Michaelis-Menten kinetics an appropriate value for T would be the time required to reduce the concentration of nutrient from c_0 to its critical concentration c_c. This can be expressed by the rate equation

$$-\frac{dc}{dt} = \frac{V_m c}{K_m + c} \tag{2}$$

which, when integrated with the boundary condition $c = c_0$ when $t = 0$ gives

$$T = \frac{1}{V_m} [K_m \ln(c_0/c_c) + (c_0 - c_c)]. \tag{3}$$

A vertical line is drawn at time T in Fig. 2 and the area to the right of this line corresponds to the proportion of circulations where the concentration of nutrient falls to a critical value before returning to the impeller or active region in the fermenter.

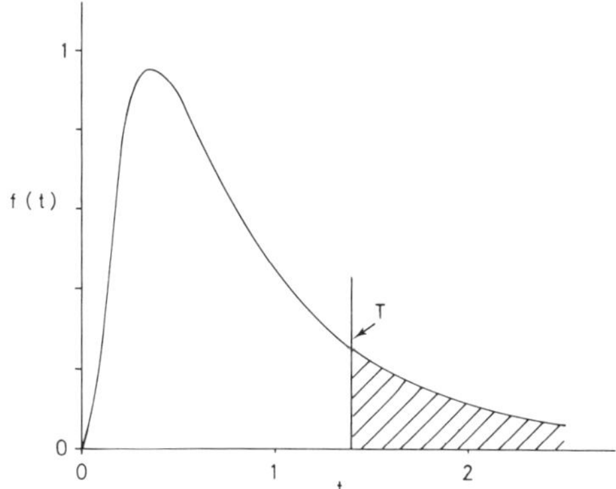

Fig. 2. Plot of distribution of circulation times

Since, in general, circulation times are log-normally distributed the shape of the curve can be described by two parameters, $\bar{t}$ and σ. The importance of both these parameters is shown in Figs. 3 and 4. In Fig. 3 two distributions are shown, both having the same value of $\bar{t}$, the mean circulation time. Curve A, having a greater standard deviation than curve B, clearly represents inferior performance of the vessel as a fermenter. Similarly in Fig. 4 two distributions are shown with equal standard deviations, but different means. Again, the importance of knowing both the mean and standard deviation is emphasised.

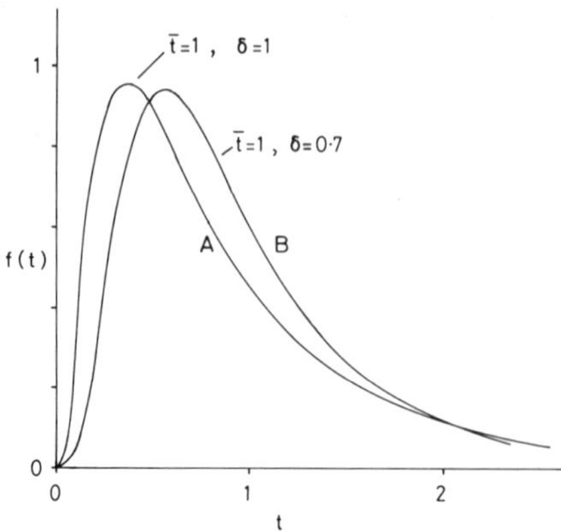

Fig. 3. Two distributions of circulation times having equal means and different standard deviations

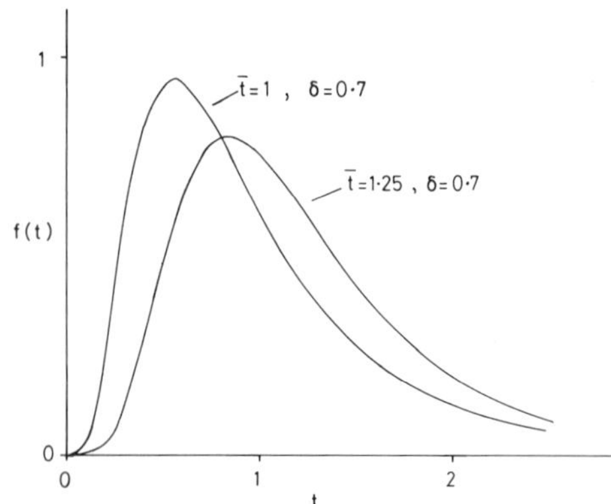

Fig. 4. Two distributions of circulation times having equal standard deviations and different means

These arguments can be put into a more mathematical form by developing a simple model of the fermentation process. The assumptions involved in the model are:

1. Oxygen, if it is the limiting nutrient, enters solution only in the region of the impeller immediately above the sparger;
2. Oxygen, or the limiting nutrient, is consumed by the microorganisms by a zero order reaction;
3. dispersion rates are negligibly small outside the active region around the sparged impeller;
4. after a period of starvation uptake of limiting nutrient proceeds at the expected rate as soon as it is made available.

The first two assumptions are conservative. If oxygen is the limiting nutrient, although it has been shown that most of the oxygen solution takes place round the sparged impeller (Wilhelm *et al.*, 1966), gas-to-liquid transfer certainly does take place in other regions of the fermenter. The second assumption corresponds to the broth in the fermenter being in a state of minimum mixedness (e.g. Levenspiel, 1962) and again this is a conservative assumption. The assumption is of course redundant if the kinetics of the reaction consuming limiting nutrient are first order.

One of the easiest parameters to calculate is the proportion of numbers of circulations whose times are equal to or greater than T. Calling this fraction L gives

$$L = \int_T^\infty f(t)dt \tag{4}$$

provided $f(t)$ has been normalised. A more useful parameter is the proportion of time spent in conditions where there is no limiting nutrient available in zero order reactions, or where its concentration is below its critical value, c_c. Calling this fraction S results in

$$S = \int_T^\infty tf(t) \cdot \left(\frac{t-T}{t}\right) dt \tag{5}$$

or

$$S = \frac{1}{2}\left[1 - \mathrm{erf}\left(\frac{\ln T/\overline{t} - \sigma_l^2/2}{\sigma_1\sqrt{2}}\right)\right] - \frac{T}{2t}\left[1 - \mathrm{erf}\left(\frac{\ln T/t + \sigma_l^2/2}{\sigma_1\sqrt{2}}\right)\right] \tag{6}$$

if $f(t)$ is a log-normal distribution such that

$$f(t)\,dt = \frac{1}{\sqrt{2\pi}\sigma_l t}\, e^{-\frac{1}{2}\left(\frac{\ln t - \mu^2}{\sigma_l}\right)} dt. \tag{7}$$

μ is the log-mean circulation time, and σ_l the corresponding standard deviation; they are related to the natural values by

$$\overline{t} = e^{(\mu + \sigma_l^2/2)} \tag{8}$$

and

$$\sigma^2 = \overline{t}^2(e^{\sigma_l^2} - 1). \tag{9}$$

Figure 5 is a plot of S against the ratio of reaction time T to the mean circulation time $\overline{t}$.

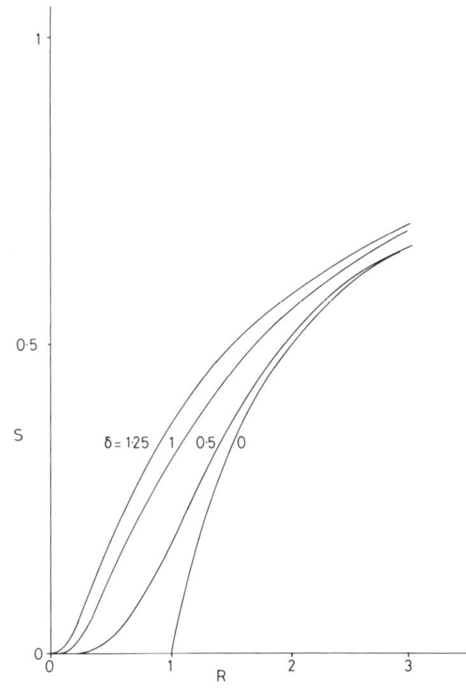

Fig. 5. Proportion of time during which no nutrient is available, S, *vs* R, the ratio t/T for different values of σ the standard deviation of circulation times

If the quality of reaction is unaffected by limiting nutrient concentration then S represents the fraction of fermenter volume wasted so far as useful fermentation is concerned. Alternatively, $(1 - S)$ is the actual yield of the fermenter compared with the yield that would have been obtained had limiting nutrient been available everywhere all the time.

Taking the point that in a zero order reaction the rate is independent of concentration, it does not necessarily follow that the actual balance of complex reactions is independent of concentration. It is possible to calculate the average and variance of the nutrient concentrations experienced by a portion of mycelium as it circulates around the fermenter. A typical portion of mycelium will experience a concentration of limiting nutrient that varies in a manner such as that shown in Fig. 6 when the reaction order is zero. Each peak on the plot represents a return to the active region where the limiting nutrient concentration is restored instantaneously to a value c_0.

If the reaction order is other than zero then the shape of the curve shown in Fig. 6 is modified to the form shown in Fig. 7.

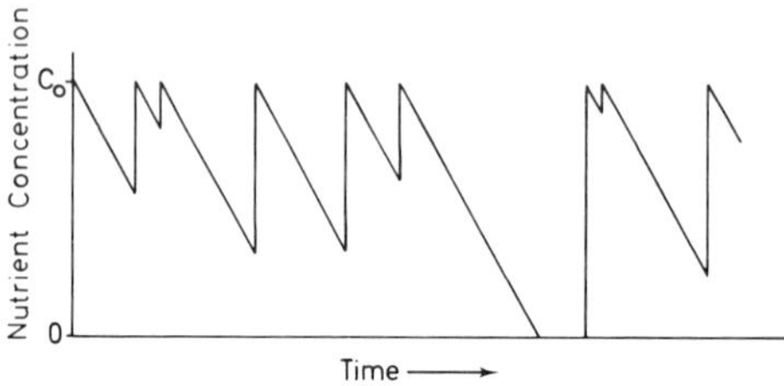

Fig. 6. Time variation of nutrient concentration of a typical portion of mycelium. Nutrient consumed by zero order reaction

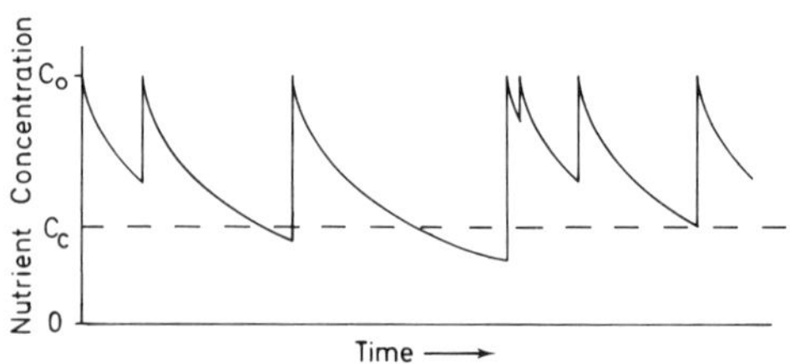

Fig. 7. Time variation of nutrient concentration of a typical portion of mycelium. Arbitrary reaction order

For a zero order reaction the mean concentrations of nutrient experienced by a portion of mycelium during a circulation of time t is

$$\bar{c} = \frac{1}{2}c_0 + (c_0 - k_0 t))$$

$$= c_0 - \frac{k_0 t}{2} \qquad \text{for } t \leqslant T \tag{10}$$

or

$$\bar{c} = \frac{c_0}{2} \cdot \frac{T}{t} \qquad \text{for } t \geqslant T. \tag{11}$$

Combining these expressions for the whole distributions of circulation times gives an overall average

$$\frac{\bar{c}}{c_0} = 1/2[1 - \mathrm{erf}(\frac{\ln R + \sigma_l^2/2}{\sigma_l\sqrt{2}})] - \frac{R}{4} e^{\sigma_l^2} [1 - \mathrm{erf}(\frac{\ln R + 3\sigma_l^2/2}{\sigma_l\sqrt{2}})]$$

$$+ \frac{1}{4R} [1 + \mathrm{erf}(\frac{\ln R - \sigma_l\sqrt{2}}{\sigma_l\sqrt{2}})]. \quad (12)$$

Figure 8 shows a plot of this dimensionless mean concentration against the reaction rate group $R = k_0\bar{t}/c_0$. By similar arguments an expression for the variance of concentration experienced by a portion of mycelium as it circulates around the tank can be found. The analytical expression is unwieldy, and the results are summarised in Fig. 9.

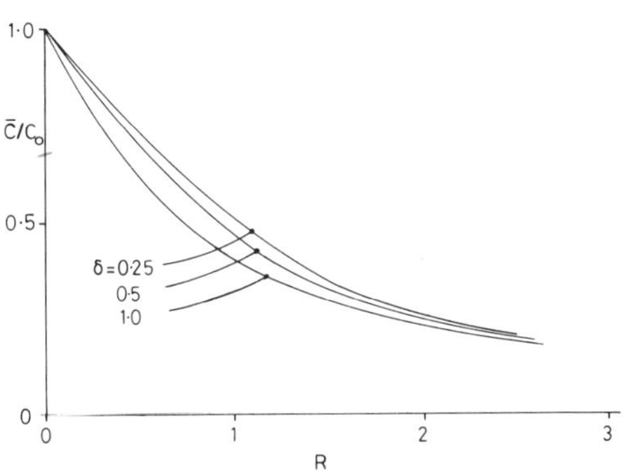

Fig. 8. Dimensionless mean nutrient concentration experienced by a portion of mycelium *vs* R, the ratio t/T for different values of σ the standard deviation of circulation times

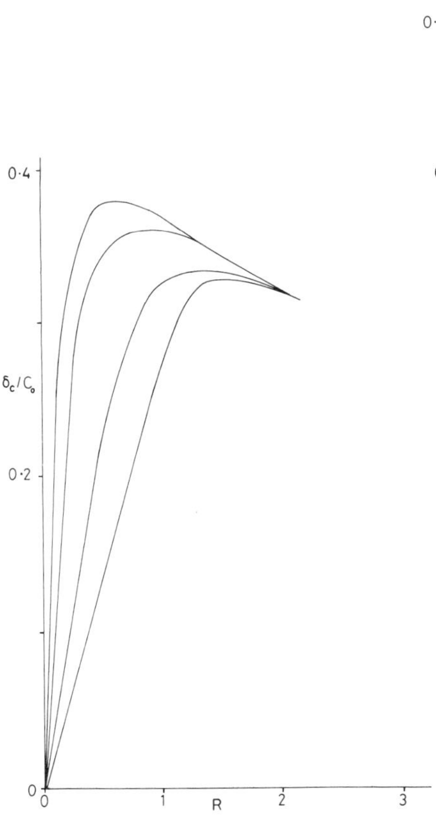

Fig. 9. Dimensionless standard deviation of nutrient concentration experienced by a portion of mycelium *vs* R, the ratio t/T for different values of σ the standard deviation of circulation times

This graph shows the ratio of standard deviation to mean concentration plotted against R for various values of σ, the standard deviation of circulation times.
The important feature of both Figs. 8 and 9 is the marked worsening in expected performance with increasing variance of circulation times. Neither figure should be taken as an exact prediction of performance, as the model involves several untested assumptions. Rather, the results should be seen as predictions of trends in performance associated with different distributions of measured circulation times.
Although zero order kinetics have been assumed there is no reason why other rate expressions should not be used instead. The exact shape of the curves will be different, but the forms will be the same, and the qualitative predictions will be unaltered.

3.2 Fermenter Design

The object of this section is to show how the basic measures of circulation, $\bar{t}$, σ, and $f(t)$, can be related to the design of the fermenter and its mode of operation. Firstly, though, the way in which these data can be treated to give an estimate of the terminal mixing time, and the effective dispersion coefficient in the system is discussed. The purpose is to show how the results obtained by the use of a flow follower can be related to other measures of quality of mixing.
A typical method of measuring terminal mixing times was referred to in the previous section. One of the difficulties of the method as described, apart from the choice of a suitable tracer, is the fact that the test is dynamic and the better the mixing the faster is uniformity attained. The same result can be found from a distribution of circulation times, $f(t)$, in the following manner.
$f(t)dt$ is, by definition, the proportion of circulations having times between t and $(t + dt)$. Therefore the proportion of a pulse of tracer returning to the impeller or active zone for the first time at time t, call it $f_1(t)$, is just

$$f_1(t) = f(t)\,dt. \tag{13}$$

The fraction returning after two circulations in time t is made up of the fraction circulating in some time θ, followed by a circulation of time $(t - \theta)$. θ is a dummy time variable in the range $0 < \theta < t$. Since there is no correlation between successive circulation times it follows that

$$f_2(t) = \int_0^t f(t - \theta) \cdot f(\theta) \cdot d\theta. \tag{14}$$

By similar convolution the fractions of original tracer returning after 3, 4 and more circulations can be found. The effective concentration of tracer at time t is then

$$r(t) = \Sigma_{n=1}^{\infty} f_n(t). \tag{15}$$

With $f(t)$ as a log-normal distribution the convolution integrals of the type shown in Eq. (14) have been summed, and Fig. 10 shows the response curves for two different values of σ, the standard deviation of $\bar{t}$, the mean circulation time. Like other recircula-

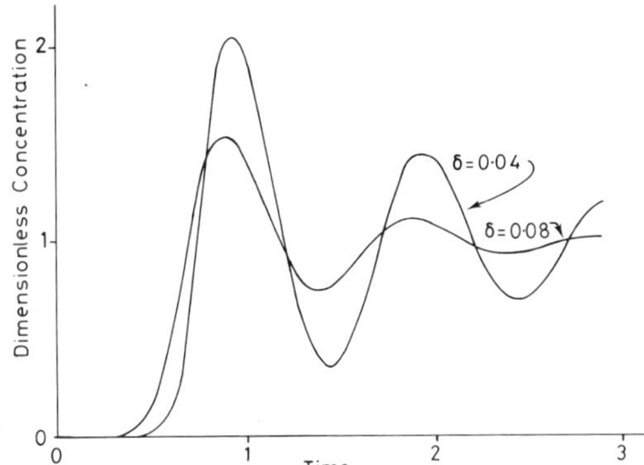

Fig. 10. Dynamic response curves for two different values of σ, the standard deviation of circulation times [Eq. (15)]

tion models (Holmes *et al.*, 1964) the limit $\sigma = 0$ corresponds to plug flow, but the other limit of perfect mixing cannot be approached with $f(t)$ in the form of a log-normal distribution. In this context the Weibull distribution is perhaps relevant as in the limits it is a simple exponential decay or a delta function corresponding to plug flow (Hahn and Shapiro, 1967) and it merits further investigation.

It is interesting to note that this analysis predicts that terminal mixing is achieved at some fixed value of the dimensionless circulation time. Since mean circulation time is inversely proportional to speed it follows that Nt_m is constant for a given system. Comparing the form of dynamic responses shown in Fig. 10 with similar responses from dispersion models leads to the conclusion that σ^2 is related to the dispersion coefficient by an equation of the form

$$\sigma^2 = k/D. \tag{16}$$

These ways of using flow follower data have been included to show the relation of flow follower data to other common measures of quality of mixing. However, the results are probably more useful in their own right.

Because of the wide variety of impeller designs, the range of possible fermenter geometry and the various forms of broth rheology, no attempt has been made, or can be made, to propose "ideal" results. What follows is an attempt to show the form of results that can be exptected, as a guide to the application of the method to particular systems. With unaerated systems the following results are found

$$\frac{V}{t} = k_1 N \tag{17}$$

and

$$\frac{V}{\sigma} = k_2 N. \tag{18}$$

Equation (17) merely expresses the well known result that pumping capacity is proportional to impeller speed. Equation (18) is more interesting, for in conjunction with

Eq. (17) it predicts that for a given system the ratio (σ/t) is constant and independent of speed.

Plots of $f(t)$ against time show that it is a very close approximation to a log-normal distribution, such that

$$f(t) \cdot dt = \frac{1}{\sqrt{2\pi}\,\sigma_l t}\, e^{-1/2\left(\frac{\ln t - \mu}{\sigma_l}\right)^2} dt.$$

Now

$$\bar{t} = e^{\mu + \sigma_l^2/2}$$

and

$$\sigma^2 = \bar{t}^2 (e^{\sigma_l^2} - 1).$$

Since the ratio (σ/t) is constant it can be defined in terms of one variable σ_l, the standard deviations of the logarithms of the circulation times.

$$\frac{\sigma}{\bar{t}} = \sqrt{e^{\sigma_l^2} - 1}. \tag{19}$$

It is worth noting that like a normal distribution a log-normal distribution is completely defined by its mean and variance.

Another experimental observation with unaerated systems is that for a given system the power input, P, is related to $\bar{t}$ and σ by

$$P \alpha \frac{1}{\bar{t}\sigma^2}. \tag{20}$$

Now a standard expression for power input in a stirred system is

$$P \alpha Q \cdot H \cdot \rho \tag{21}$$

where Q = pumping capacity
H = head
ρ = density of fluid.

Clearly from Eq. (17)

$$Q \alpha 1/\bar{t}$$

and so for a liquid of constant density the variance of circulation times can be interpreted as a head. Variance has already been related to dispersion coefficient in Eq. (16). Returning to the basic Eqs. (17) and (18), the problem is to define more closely the constants of proportionality in terms of known geometry. Equation (17) can be extended to

$$\frac{V}{\bar{t}} = k_3 N D^3 \tag{22}$$

and similarly Eq. (18) becomes

$$\frac{V}{\sigma} = k_4 N D. \tag{23}$$

Further experimental work would be required to define the constants k_3 and k_4 more completely until a stage was reached where each was a characteristic of a particular impeller design, analogous to the constants in Rushton's power equations.
As well as being functions of impeller geometry the constants k_3 and k_4 depend on the flow properties of the fluid being stirred.
Preliminary analysis of results obtained with Newtonian liquids of different viscosities suggests that

$$k_3 \text{ and } k_4 \ \alpha \ (a + b\mu)$$

where a and b are constants and μ is the viscosity. Furthermore it is known that the position of the impeller in a tank affects the values of the constants as indeed does the geometry of the tank. Not enough is yet known to allow functional relations to be established. In unaerated systems it is then possible to relate the basic parameters $\bar{t}$ and σ both to design variables such as impeller type and size and to fluid properties.
The situation with aerated systems is that a radio flow-follower is equally satisfactory. The simple relations given by Eqs. (17) and (18) no longer apply. Neither $\bar{t}$ nor σ is related simply to impeller speed N, but they are related to each other, when the impeller is operated under "flooded" conditions, by an equation of the form

$$\bar{t} \alpha \sigma^n \tag{24}$$

where n is a constant, and $n \neq 1$.
Similarly there is no simple relation between power input and $\bar{t}$ and σ in aerated systems, but under conditions of flooding the following form of relation was observed

$$\left(\frac{1}{\bar{t}\,\sigma^2}\right)_0 - \left(\frac{1}{\bar{t}\,\sigma^2}\right)_a \alpha N^3 \tag{25}$$

where the subscript 0 refers to values obtained in an unaerated system and a refers to the aerated system. It will be noted that the air flow rate does not enter into this equation. In general the distribution of circulation times is again well approximated by a log-normal.

The addition of air to a system naturally introduces more variables into an already complex system. Further work is obviously required in order to establish the values of the constants of proportionality for particular systems.

4. Discussion

The whole basis of the method involving the use of a flow-follower is that the frequency with which broth passes through a known region in the fermenter can be determined accurately and automatically. The use of a flow-follower incorporating a radio transmitter is particularly convenient and meets the prerequisites suggested in Section 2. This article has considered in particular those cases in which the active zone is the region closely surrounding the sparged impeller, with the implication that oxygen supply is the critical function of agitation. However, there is no reason to limit the application of the technique to fermentations where oxygen supply is limiting or to this particular zone or aeration/agitation system. For example if the limiting nutrient were supplied as an aqueous solution then the active zone would be at the point of addition, and the critical function of agitation would be to distribute this nutrient as fast as possible.

It has been assumed that once an element of broth has been recharged with limiting nutrient to some high concentration, c_0, then reaction can proceed without hindrance until the element returns to the active region, or until nutrient is depleted. An extension of this view is to consider the size of this element of broth. One of the fundamental ideas of turbulence theory is that there is a minimum eddy size which will be produced in any particular hydrodynamic situation, and so all transfer over distances less than this scale must rely on molecular diffusion. Although this eddy size cannot be measured in a fermenting broth, the ensuing analysis may be applied to the interpretation of performance data.

For a liquid of fixed density and viscosity, the minimum eddy size is related to the specific power input in the form given below

$$R \propto (P/V)^{-1/4}. \tag{26}$$

Now suppose that such eddies have identities and exist, then in effect there is a spherical element of broth of radius R whose interior relies solely on molecular diffusion for the supply of nutrients from some external fluid of nutrient concentration c_b. Similarly, products of the fermentation can only leave the eddy by diffusion. This situation is identical to that of a conventional spherical catalyst particle, and the concept of an effectiveness factor may be used (Satterfield and Sherwood, 1963). The effectiveness factor is defined as the ratio of the amount of reaction that takes place in the particle, to the amount that would have taken place if the whole sphere had had nutrient available at concentration c_b.

Consider the sphere shown in Fig. 11, and a spherical shell of radius r and thickness dr. In a steady state the rate of diffusion into the shell at radius $(r + dr)$ must equal the rate

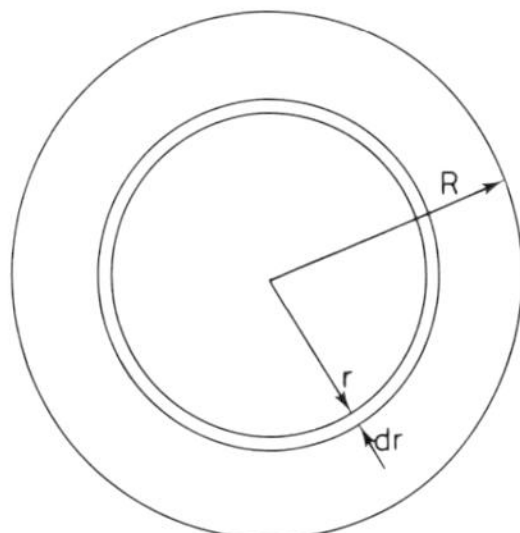

Fig. 11. Section through spherical eddy of radius R

of diffusion out at radius r towards the centre plus the rate of consumption in the volume of the shell. This may be expressed as

$$4\pi(r + dr)^2 \cdot D\left(\frac{dc}{dr} + \frac{d^2c}{dr^2} \cdot dr\right) - 4\pi r^2 \cdot D \cdot \frac{dc}{dr} = 4\pi r^2 dr \cdot k'_0 \tag{27}$$

where D is the effective diffusivity of nutrient through the broth forming the eddy, and k'_0 is a volumetric zero order rate constant. Equation 27 may be simplified to give

$$\frac{d^2c}{dr^2} = \frac{2}{r} \cdot \frac{dc}{dr} = \frac{k'_0}{D}. \tag{28}$$

The two boundary conditions necessary to solve Eq. (28) are

$$c = c_b \text{ at } r = R$$

and that c is finite or zero at $r = 0$.
The solution of Eq. (28) is then

$$c = c_b - \frac{k'_0}{6D}(R^2 - r^2). \tag{29}$$

Since the reaction is zero order the concentration c can take a value of zero, and the first step now is to calculate the maximum eddy radius, R_m, such that all the broth within it is supplied with nutrient. Putting both c and r equal to zero in Eq. (29) gives

$$R_m = \sqrt{\frac{6DC_b}{k'_0}}. \tag{30}$$

If $R > R_m$ then there is a region in the centre of the eddy that must be starved of nutrient. Calling the radius of this region R_c gives

$$R_c = \sqrt{R^2 - \frac{6Dc_b}{k'_0}}. \tag{31}$$

Clearly the ratio of volume of eddy in which nutrient is available to the total volume of the eddy is the effectiveness factor, η,

$$\eta = \frac{R^3 - R_c^3}{R^3}$$

$$= 1 - \left(\frac{R_c}{R}\right)^3 . \tag{32}$$

Provided $R \leqslant R_m$ then the effectiveness factor is unity, otherwise Eq. (32) is used. Figure 12 shows a plot of the form of relation between η and the group $R/\sqrt{\frac{k_0'}{6Dc_b}}$. But R is related to specific power input and combining Eqs. (26) and (32) gives a plot of the form shown in Fig. 13.

Although this analysis rests on a number of simplifying assumptions it is suggested that if performance is related to power input in the form shown in Fig. 13 some interesting implications may be drawn. One obvious refinement is to relax the steady-state assumption used in the derivation of an effectiveness factor and to assume instead that at time

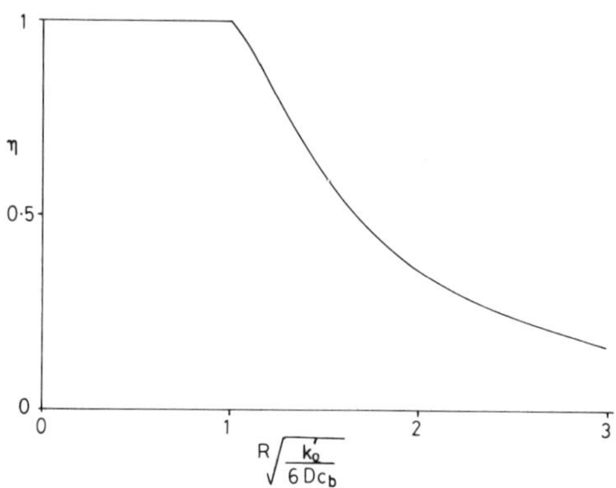

Fig. 12. Effectiveness factor *vs* reaction diffusion modulus

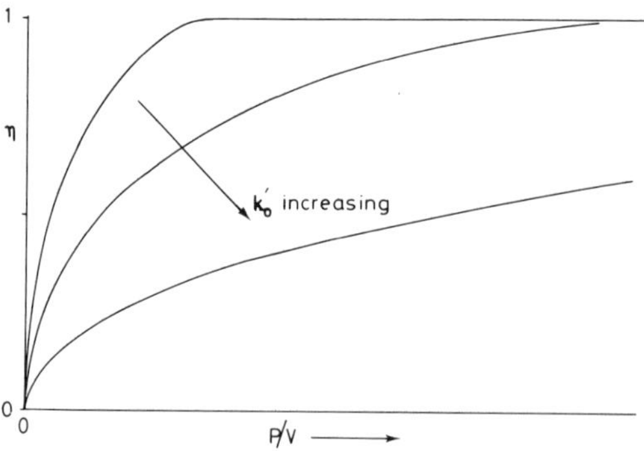

Fig. 13. Effectiveness factor *vs* specific power input for different values of the zero order rate constant

0 the concentration everywhere in the eddy is c_b and then to use an unsteady-state approach. This model could be used in conjunction with the circulation model described in Section 3.1, but it is felt that unless and until there are clear indications that such an approach is desirable work on it would have little real value.

Turning now to a more general discussion of the model the assumption of constant viscosity used in Eq. (26) can be relaxed to give a more complete relation of the form

$$R \alpha (P/V)^{-1/4} (\mu)^{3/4} (\rho)^{-1/2}. \tag{33}$$

In practice it is not possible to vary the density of a broth, but by operating with different concentrations of biomass the viscosity can be changed. For convenience the Arrhenius relationship between the concentration of suspended solids and viscosity will be assumed.

$$\mu = \mu_0 e^{\alpha x} \tag{34}$$

where α is a constant such that $\alpha > 0$, μ_0 is the viscosity of the filtered broth, and x is the biomass concentration.

The productivity of a fermenter is given by

$$\frac{Q}{V} = \eta \cdot k_0' \tag{35}$$

assuming that reaction rate is proportional to productivity, and assuming further that the rate constant k_0' is itself proportional to biomass concentration, say $k_0' = hx$, leads to

$$\frac{Q}{V} = \eta h x. \tag{36}$$

Combining Eqs. (36, 34, and 33) gives an overall expression for productivity in terms of (P/V) and biomass concentration of the form

$$\frac{Q}{V} = h x \left[1 - \left(\frac{\sqrt{(P/V)^{-1/2} (\mu_0 e^{\alpha x})^{1 \cdot 5} - \frac{6Dc_b}{nx}}}{(P/V)^{-1/4} (\mu_0 e^{\alpha x})^{3/4}} \right)^3 \right]. \tag{37}$$

This equation predicts that for constant power input, productivity is related to biomass concentration in the manner shown in Fig. 14. The maximum productivity is reached when the radius of the eddy equals R_m, defined in Eq. (30). It can be seen that to maintain this condition the specific power input must rise steeply with biomass concentration as is shown in Fig. 15.

There is no reason why other rate expressions should not be used in plaxe of the zero order kinetics used in this model. A very complete coverage of the methods of calculating effectiveness factors for the non-linear kinetics typical of fermentation has been presented (Moo-Young and Kobayashi, 1972). It is worth noting in this connection that effectiveness factors are not necessarily less than unity if the reaction is subject to inhibition.

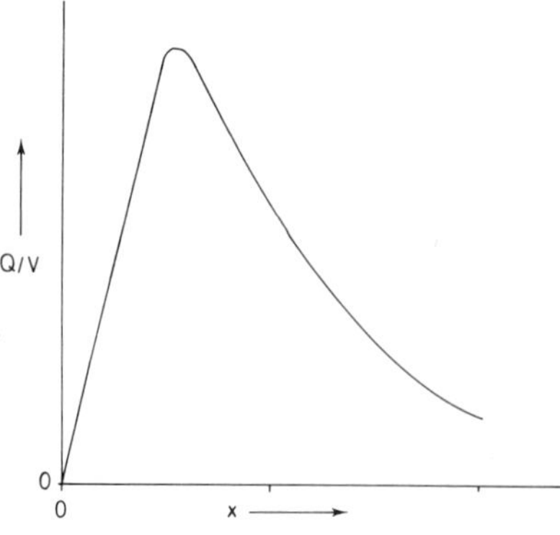

Fig. 14. Productivity, Q/V *vs* biomass concentration

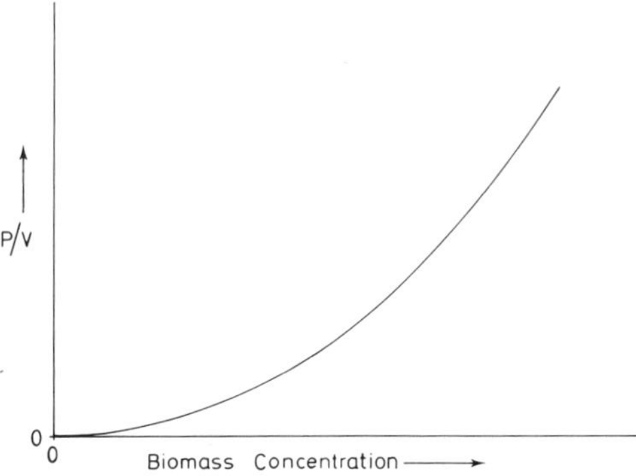

Fig. 15. Specific power input required to maintain effectiveness factor at unity *vs* biomass concentration

Again it is stressed that although these analyses rest on a number of simplifying assumptions they do allow some interesting implications to be drawn about the relationship between performance and specific power input and broth viscosity or biomass concentration.

Appendix: Radio Flow-follower

In principle any flow-follower can be used, provided that it is approximately neutrally buoyant in the fermentation medium. In practice limitations arise: for example direct visual observation is not generally possible. The adequacy of the characterization is improved as the size of the follower decreases, but this increases the difficulties and reliability of detection. The characterization described here was obtained by the use of a small radio-transmitter which meets all the operational prerequisites mentioned above, and it limited only by the life of the power source.

The most important part of the equipment is the radio flow-follower, for which a circuit diagram is shown in Fig. 16. It is housed in a plastic sphere, 2 cm diameter, made

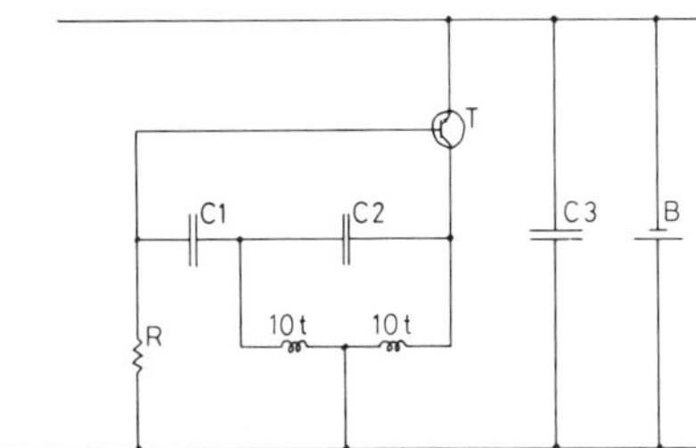

Fig. 16. Circuit diagram of radio transmitter components. B Mercury cell 1.35 v. C_1 0.01 μF. C_2 47 pF. C_3 2 μF. R 100 KΩ. T Bc 112 (Bc 146)

as two hemispheres which can be screwed together, thus allowing the battery to be changed easily and facilitating buoyancy adjustment.

Although the follower will not withstand steam sterilisation, it can be chemically sterilised for aseptic addition to a fermenter. A small pipette tank with wide-bore valves is suitable for the purpose.

The only other piece of equipment that has to be used inside the fermenter is the receiving aerial, in the form of a loop closely surrounding the impeller or active region. The lead from the aerial can be taken out through the lid of the fermenter. A stub aerial is required so that the lead inside the fermenter does not pick up signals when the follower passes close to it.

The receiver should have a continuous frequency range so that it can be tuned to the frequency of the transmitter. Typically the circuit shown in Fig. 6 has a frequency of 10 MHz.

The output from the receiver is amplified and used to trigger a timer circuit to record the passage of the follower through the active region. The data logger should also display elapsed time, the number of events and the time at which the last passage accured. The first two displays allow a ready check on the mean circulation time at any stage of the experiment, and the third is useful if there is any doubt about the correct functioning of the follower. Figure 17 shows a block diagram of the equipment.

To use the equipment a battery is fitted to the follower and its buoyancy is adjusted. The frequency of the receiver is adjusted to that of the transmitter and the system is ready for the final adjustment of the trigger circuit. The requirement is that when the follower passes through the active region the signal strength should be just enough to

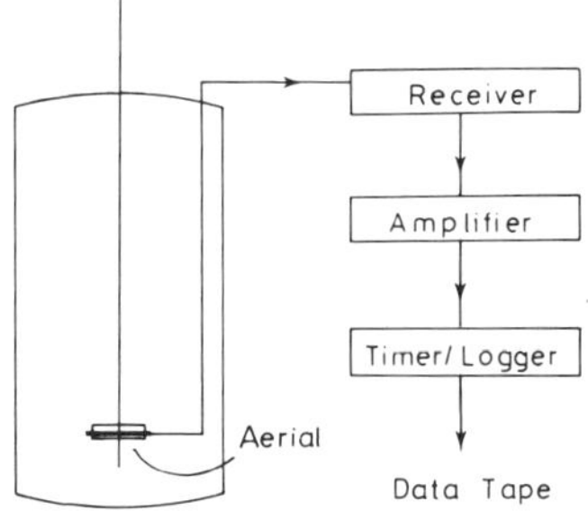

Fig. 17. Layout of equipment for radio flow-follower experiments

activate the logger. The sketch in Fig. 18 shows how the follower should be held for this adjustment when an impeller is employed. Any upward vertical movement of the follower from the position shown will cause a reduction in signal strength such that the logger relay no longer closes. When working with large vessels, or in aseptic conditons this adjustment can be made outside the vessel using a test aerial of the same characteristics as that inside.

When these tests have been completed satisfactorily the equipment is ready for use. The exact details of the radio receiving equipment and the data logger are not critical, and for that reason only an indication of what is necessary has been given: availability of apparatus may well determine the actual design.

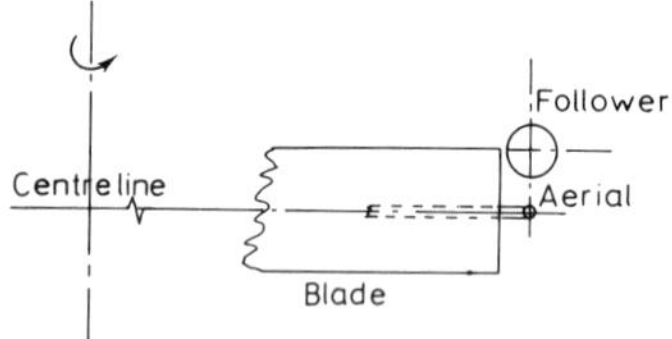

Fig. 18. Position of follower relative to aerial and impeller blade when setting gain of receiver circuit

Nomenclature

a, b	constants
c	concentration of limiting nutrient
c_b	bulk concentration of limiting nutrient
c_c	critical concentration of limiting nutrient
c_0	concentration of limiting nutrient in active region
$\overline{c}$	mean concentration of limiting nutrient
D	impeller diameter
or	effective diffusivity [Eqs. (27), (28), (30), and (31)]
$f(t)$	distribution of circulation times
$f_n(t)\,dt$	proportion of pulse input circulating n times in time t
h	constant
H	impeller head
k, k_1, k_2, k_3, and k_4	constants
k_0	zero order reaction rate constant
k_0'	volumetric zero order rate constant
K_m	Michaelis constant
L	fraction of circulations having times $\geqslant T$ [Eq. (4)]
N	impeller speed
n	constant
P	power input
Q	impeller pumping capacity
r	radius in eddy
$r(t)$	concentration of pulse of tracer at time t
R	reaction rate group = $k_0 t/c_0$
or	eddy radius [Eq. (26) *et seq*]

R_c	radius of core of eddy starved of nutrient
R_m	maximum radius of eddy in which nutrient is everywhere available [Eq. (30)]
S	fraction of time spent in starvation conditions
t	time
$\bar{t}$	mean circulation time
t_m	terminal mixing time
T	characteristic reaction time defined by Eqs. (1) and (3)
V	volume of fluid
V_m	maximum reaction velocity
x	biomass concentration
α	constant
η	effectiveness factor
θ	dummy time variable
μ	log mean circulation time
	or viscosity in Eq. (33)
ρ	fluid density
σ	standard deviation of circulation times
σ_l	standard deviation of log circulation times

References

Aiba, S.: A. I. Ch. E. Journal **4**, 485 (1958).
Aiba, S., Humphrey, A. E., Millis, N. F.: "Biochemical Engineering", 2nd Edn. (1973).
Blakebrough, N., Sambamurthy, K.: Biotechnol. Bioeng **7**, 25 (1966).
Finn, R. K.: in "Biochemical and Biological Engineering Science", Ed. N. Blakebrough. London: Academic Press (1967).
Hahn, G. J., Shapiro, S. S.: "Statistical Models in Engineering". London: John Wiley & Sons Inc., 1967.
Holmes, D. B., Voncken, R. M., Dekker, J. A.: Chem. Engng Sci. **19**, 201 (1964).
Levenspiel, O.: "Chemical Reaction Engineering". New York & London: John Wiley & Sons, Inc., 1962.
McManamey, W. J., Loucaides, R., Lewis, M. J.: Biotechnol Bioengng, Symp. **4**, 379 (1973).
Metzner, A. B., Taylor, J. S.: A. I. Ch. E. Journal **6**, 109 (1960).
Moo-Young, M., Kobayashi, T.: Can. J. Chem. Engng. **50**, 162 (1972).
Riet, K. van't: "Turbine Agitation Hydrodynamics and Dispersion Performance". Ph. D. Thesis, Delft University (1975).
Satterfield, C. N., Sherwood, T. K.: "The role of diffusion in catalysis". London: Addison Wesley Publishing Co Inc. (1963).
Sykes, P.: Chem. Engng. Sci. **20**, 1145 (1965).
Uhl, V. W., Gray, J. B.: "Mixing–Theory and Practice". New York: Academic Press (1966).
Wilhelm, R. H., Donohue, W. A., Valesano, D. J.: Biotechnol. Bioengng **8**, 55 (1966).

The Immobilization of Whole Cells

T. R. JACK
Scarborough College, University of Toronto, West Hill, Ontario

J. E. ZAJIC
Faculty of Engineering Science, The University of Western Ontario, London, Ontario

Contents

1. Introduction

1.1 The Advantages

The attachment of enzymes to an inert matrix is an important and well-developed procedure for both practical and academic purposes. The enzymes employed are often of microbial origin and an apparent extension of the immobilization concept is the fixation of the whole microbial cell to an inert support without the prior separation and purification of the particular metabolic components of interest. Over the last few years, sufficient information on this approach has appeared in the literature to evaluate its potential in a wide variety of scientific and industrial applications.

The advantages of immobilizing *the entire cell* rather than a purified enzyme are numerous: the expense of separation, isolation and purification of the enzyme is obviated; a wider scope of reactions is possible including multi-step reactions utilizing several enzymes; maintainence of the enzyme in its native state enchances its stability; and the presence of co-factors and continued biosynthesis within the cell contribute to the longevity of enzymatic activity. Further, the application of immobilized whole cells need not be limited to the production of classical fermentation products but can be expanded to include the production of viral particles or synchronous cells, the chromatographic separation of special cells and the culturing of animal tissue.

Fixation of enzymes on an inert support either as whole cells or in a purified form yields many advantages over the use of free cells in classical fermentation processes. Tabulations of these operational advantages have been published elsewhere [1], but include: the possiblity of eliminating feedback inhibition or consumption of product by the use of novel column or surface reactor designs which allow the articulate manipulation of reaction conditions and swift removal of products, the operation of processes in a continuous fashion and the feasibility of conducting sequential reactions in a chain of connecting continuous bioreactors.

It is the purpose of this review to illustrate the exploitation of these advantages based on material already published in the literature and to collect information relevant to the techniques of whole cell immobilization from diverse areas of publication.

1.2 Historical Perspective

The affinity of certain microorganisms for growth on surfaces is a well established phenomenon [2]. Under natural conditions, microbial films develop on a wide variety of biotic and abiotic supports including such diverse materials as sand grains, teeth, intestinal villi, polyvinylchloride tubing, mineral faces and metal surfaces.

In many examples, this adsorption arises out of physical effects such as ionic attraction while in others a more active role is played by the microorganism such as is seen in the prosthecate bacteria *Caulobacter* which form an extracellular adhesive disc or holdfast to anchor the cell to solid surfaces.

Zvyagintsev *et al.* [3] have actually measured the magnitude of the adhesive force of a variety of microbes on glass and report a large species specific range of strengths of

attachment. Interestingly, for some microbes such as the genus *Bacillus,* the proportion of cells adhering is small but those which do adhere are very firmly held and conversely, for the genus *Pseudomonas*, the cells adhere in large numbers via a weak interaction. It would seem, therefore, that for such physical adsorption phenomena, the number of cells adsorbed does not correlate with the strength of the adsorption interaction.
In certain classical fermentations (see Table 1), the microbes are purposely provided with a surface to enhance their growth and, in a broad sense, are immobilized in this fashion. In a continuous reaction vessel, the trivial wash out state often cannot be attained [5] due to the adherence of some part of the microbial population to the internal surfaces of the reactor itself.

Table 1. Microbial processes using biological films ([4], reproduced with permission from Biotech. and Bioeng.)

Process	Objective	General characters
Biological wastewater treatment		
Trickling filter	Biological oxidation of industrial and domestic effluent	Nonaseptic, microbial growth occurs in a packed bed. Wastewater distributed intermittently over the packing. Aerobic; packing supported on a grid structure, enhancing aeration by natural convection
Rotating disc	Biological oxidation of industrial and domestic effluent	Microbial growth on discs rotating in a vertical plane, the discs dipping into a trough of wastewater. Microbial growth is alternately in contact with nutrients and air
"Quick" vinegar process	Oxidation of alcohol by acetic acid bacteria	Similar in principle to the trickling filter, but with forced aeration. Wine or other feed liquor recirculated over beechwood chips or similar packing. Batch process (4–5 days)
Animal tissue culture	Growth of animal cells in a surface layer for the culture of viruses	Animal tissue minced and reduced to single cells by enzyme action. The cells adhere to surfaces provided and grow as a film in the presence of a suitable medium. Can be used subsequently for virus culture. Strictly aseptic
Bacterial leaching of ores	Recovery of metals from sulfide ores using iron and sulfur oxidizing bacteria	Bacteria used, *in situ* in dumps of low-grade or waste ores. Possibility of tank-leaching methods

Animal tissue cultures of course, are routinely grown as a surface monolayer. The extensive literature on this topic and its attendent area of virus production from surface bound cells has been reviewed recently and will not be dealt with here [6].
Thus, the immobilization of whole cells is not a novel concept, but rather a refinement of a phenomenon observed in nature and in those industrial microbiological processes in which the surface growth of cells is favoured.

2. Methods of Immobilizing Whole Cells

The methods of whole cell immobilization which have been reported may be categorized as follows:
(a) Entrapment by an inert support
(b) Adsorption by an inert support
(c) Binding via immobilized biological macromolecules
(d) Covalent or coordinate bonding of the cell to an otherwise inert support.
Subsequent discussion of the applications of immobilized whole cells has been subdivided according to the immobilization technique employed since the criteria imposed by the immobilization technique usually determine the nature of the application. For each case, an effort has been made to indicate the scope of the technique in terms of the microorganisms studied, the relationship of the physiological state to activity, the comparable activity of free cells, the stability of the immobilized system and the degree of cell detachment (especially of new cells) encountered during use.

2.1 Entrapment Techniques

In applications primarily concerned with the enzymatic conversion of a water soluble substrate to product, the advantages gained by whole cell immobilization can be achieved simply by containing the cells in or on a second water insoluble phase.

2.1.1 Polyacrylamide Gels

Method
The most frequently employed technique is the entrapment of whole cells in a polyacrylamide gel.
The general procedure involves the polymerization of an aqueous solution of acrylamide monomers in which the microorganisms are suspended. The technique is straightforward and generally results in an effective entrapment of the cells with limited modification of the cellular enzymes.
The cell-containing polymeric gel so obtained can be easily granulated for use as a column packing, the porosity of which is a function of the degree of cross-linking in the acrylamide itself.
The polymerization of acrylamide takes place by a free radical process in which linear chains of polyacrylamide are built up. The inclusion of a bifunctional reagent which has two unsaturated double bonds susceptible to inclusion in the polymer results in cross-links between the polymer chains. The degree of cross-linking then is a function of the relative amounts of acrylamide monomer and bifunctional cross-linking agent.
Of the various cross-linking reagents tested, N, N′-methylenebis(acrylamide) is preferred on the basis of the entrapped cell enzyme activity and the physical properties of the gel obtained [7].
The relative concentrations of cells, acrylamide and N,N′-methylenebis(acrylamide) in the polymerization solution can influence the enzyme activity of the entrapped cells.

Table 2 shows the optimal solution compositions found in three studies. These are maximized with respect to the enzyme activity noted.

Table 2. The composition of the polymerization solution optimized for the maximum enzyme activity of the polyacrylamide entrapped cells

Microorganism	Assayed activity	Solution composition by weight			References
		Cells	Acrylamide	N,N'-Methylenebis (Acrylamide)	
Brevibacterium ammoniagenes IFO 12071	coenzyme A production	15–20%	15%	0.8%	Shimizu *et al.* [8]
Escherichia freundii K-1	p-nitrophenyl phosphate hydrolysis	6%	15%	0.4%	Saif *et al.* [19]
Gluconobacter melanogenus	L-sorbosone production	5%	15%	0.8%	Martin *et al.* [9]

The free radical polymerization reaction can be initiated by either a chemical or photochemical catalyst system under anoxic conditions.
The most common catalyst systems consist of the persulfate ion and β-dimethylaminopropionitrile or N,N,N′,N′-tetramethylethylenediamine for a chemical catalyst system and riboflavin, sodium hydrosulphite and N,N,N′,N′-tetramethylethylenediamine for a photochemical one.
The temperature at which the polymerization is carried out can also influence the activity of the immobilized cells. Generally, the lower this temperature the less the damage incurred by the microorganisms [8, 9]. For example, exposure of *Gluconobacter melanogenus* to be a solution containing 60 mg/ml acrylamide and 12 mg/ml N,N′-methylenebis(acrylamide) results in no significant loss of viability after 5 min at 15° C but total inviability after just 2 min at 45° C [9]. Similarly, the ability of gel entrapped *Brevibacterium ammoniagenes* to produce coenzyme A falls off with increasing temperature of polymerization; 100% at 0° C, 85% at 20° C and 75% at 37° C. In some cases, however, the damage incurred by the cell may actually enhance the apparent enzyme activity. This is presumably accomplished by reducing the barrier presented by the cell membrane [9]. Under these circumstances, the temperature for the polymerization may be optimized relative to the desired enzyme activity.
Usually the polymerization is carried out in a isotonic solution buffered with phosphate to minimize damage to the cellular machinery.

Early Demonstrations
The first papers reporting the entrapment of whole cells by polyacrlyamide gels were concerned with demonstrating that the cells so immobilized could retain viability and enzyme activity.

In a 1966 paper, Mosbach and Mosbach [10] mentioned that air dried, powdered thalli of the lichen, *Umbilicaria pustulata,* entrapped in this way retained esterase and decarboxylase activity in periodic tests over a three month period at 20° C.

Entrapped cells of the protozoa, *Tetrahymena pyriformis* and of the bacteria, *Escherichia coli* were shown by several methods to retain their viability. The uptake of glucose and oxygen across a column packed with the cell-containing gel was observed. The protozoa were observed under microscope examination to be actively struggling within their polymer cages for up to five days and the viable cells excluded 0.1% trypan blue dye. For cells of *E. coli* where motion in the gel was indistinguishable from Brownian motion, the entrapped cells were found to produce colonies on agar when freed from the gel particles by a non-destructive grinding operation [11].

The first actual application of polyacrylamide entrapped cells was in the sterospecific 11-β-hydroxylation of a steroid (Reichstein compound S) to form cortisol in a batch process [12].

The enzyme activity, provided by gel entrapped mycelia of the fungus *Curvularia lunata,* fell to about 50% after one week storage at 4° C but could be regenerated and actually enhanced by incubation of the gel in a nutrient medium containing the substrate steroid. Whether hydroxylase activity was reinduced by this treatment or whether some growth of the fungus had occurred was not distinguished. The gel immobilization of the mycelia did avoid the often incurred difficulties inherent in separating the transformed steroid from the cell material. The porosity of the gel support in this study proved sufficient to allow free passage of large steroid substrates and products to and from the entrapped mycelia [12]. As will be seen subsequently, other entrapment techniques often impose a diffusion barrier between the substrate solution and the contained cells.

Franks [13, 14] was able to study the products released from arginine catabolism (Fig. 1) by passing a phosphate buffered arginine solution through a column packed with cells of *Streptococcus faecalis* ATCC 8043 in polyacrylamide gel. The concentrations of arginine, citrulline and ornithine were monitored in the column effluent.

The L-arginine metabolizing ability of the immobilized whole cell preparation was stable over eleven days as was the integrity of the cells within the gel as shown by electron microscopy. Disruption of the cells with lysozyme allowed the diffusion of cell materials from the gel to leave a polymer matrix void of arginine catabolic activity (Fig. 1).

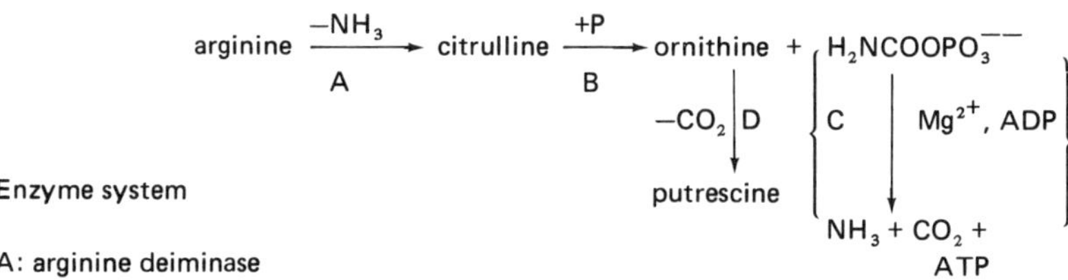

Enzyme system

A: arginine deiminase
B: ornithine transcarbamylase
C: carbamylkinase
D: ornithine decarboxylase

Fig. 1. The L-argine catabolic pathway

Several observations were made that are of general interest in the use of immobilized whole cells for multi-enzyme conversions.

First, it was observed that ornithine was the major product released from arginine catabolism. The intermediacy of citrulline in the process as shown in Fig. 1 was supported by the fact that, when previously frozen cells were immobilized, citrulline was the major product (> 90%) of arginine catabolism. Taken with the observation that exogenous citrulline was not metabolized by immobilized intact cells, it was suggested that the cell membrane, if intact, presented a barrier to citrulline diffusion. Franks also suggested that for the whole cells a spatial organization of the enzymes arginine deiminase and ornithine transcarbamylase exists such that the intermediate citrulline is directly converted to ornithine with the cofactors Mg^{2+} and ADP being provided within the cells. The subsequent conversion of ornithine to putrescine was effected only on prolonged incubation of the cells. The lack of putrescine produced by the column reactor could be attributable to the high pH (7.1), the age of the cells selected for immobilization, the selectivity of the cell membrane and/or the relative kinetic rates of conversion.

In summary, polyacrylamide entrapped microbial cells were used to carry out a multistep conversion of L-arginine to ornithine by a continuous column process without the presence of significant amounts of intermediate or subsequent products (i.e., citrulline or putrescine).

Applications

Industrial applications of the polyacrylamide entrapment technique began to appear in 1973. In a continuing program of applied research by Sato, Chibata, Tosa *et al.*, polyacrylamide immobilized cells have been the basis for the continuous production of urocanic acid [15], L-citrulline [16] and L-aspartic acid [17, 18, 7]. The L-aspartic acid synthesis has been tested on a commerical production level.

The L-citrulline production scheme is based on the metabolism of L-arginine as shown in Fig. 1.

The L-citrulline was obtained from L-arginine using gel-entrapped cells of *Pseudomonas putida* ATCC 4359 which retain 56% of their L-arginine deiminase activity on immobilization. Encasement of the cells in polyacrylamide gel in this case resulted in a shift of the temperature optimum of the enzyme activity from 37° C to 50° C. The continuous operation of a column fed L-arginine hydrochloride (0.5 M) at 37° C, pH 6.0 with superficial velocity of $\leqslant 0.26$ gave a quantitative conversion of the substrate to product for over three weeks. The recovery of crystalline L-citrulline from the effluent was accomplished in >96% overall yield [16]. Urocanic acid, a sun screening agent of pharmaceutical and cosmetic importance, can be obtained from L-histidine by the action of microbial L-histidine ammonia-lyase. Comparison of the activity of this enzyme before and after immobilization of the cells in polyacrylamide gel showed that, of the microbes tested, *Achromobacter liquidum* IAM/1667 retained its enzyme activity to the greatest degree (Table 3). It is also of note that the polyacrylamide gel procedure can have an unpredictable effect on the whole cell enzyme activity. This ranges from total inactivation as seen for *Micrococcus ureae*, Table 3, to enhancement as noted later for *Corynebacterium glutamicum.* The employment of *Achromobacter liquidum* in urocanic acid production necessitated the selective inactivation of its urocanase enzyme system which converts the desired

Table 3. Immobilization of several microorganisms having histidine ammonia-lyase activity ([15], reproduced with permission from Biotech. and Bioeng.)

Microorganisms	Enzyme activity (μmol/hr/mo)		Yield of enzyme activity (%)
	Intact cells	Immobilized cells	
Achromobacter aceris	3.4	0.7	21.6
A. liquidum	34.2	21.8	63.8
Agrobacterium radiobacter	23.2	12.0	51.7
A. tumefaciens	18.5	9.2	49.8
Flavobacterium flavescens	31.0	18.7	60.3
Micrococcus ureae	107.5	0.3	0.3
Sarcina lutea	14.0	4.7	33.6

product to imidazolone propionic acid. This was accomplished without detriment to the desired activity by heating the cells at 70° C for thirty minutes before immobilization [15].

Production was carried out in shake flasks at 37° C for one hour in a 0.25 M solution of L-histidine. The free cell activity was greatly enhanced by the addition of 0.025% of an amphipathic surface agent, cetyltrimethylammonium bromide, which was thought to facilitate the passage of substrate and product through the cell membrane. This dependence vanished on immobilization of the cells suggesting to the authors that the polymerization procedure had in some way increased the cell permeability [15].

In continuous operation in a column, the enzyme activity began to diminish after about five days operation unless the feedstock was supplemented with Ca^{2+}, Co^{2+}, Mg^{2+}, or Zn^{2+}. Addition of Mn^{2+} or EDTA accelerated this depletion of L-histidine ammonia-lyase activity. Since this enzyme is known to require Mg^{2+} to maintain activity, this deactivation was attributed to the gradual leaching of metal ions from the immobilized cells.

This may be a further manifestation of the increased permeability of the cell membrane. With Mg^{2+} present (1 mM), a feedstock of 0.25 M L-histidine at pH 9.0 and superficial velocity = 0.06 gave 100% conversions to urocanic acid in a continuous column process for over 40 days at 37° C [15].

The most thoroughly developed of these systems is based on the conversion of ammonium fumarate to L-aspartic acid by the biocatalysis of entrapped *Escherichia coli* cells. The retention of aspartase activity was screened in a variety of immobilized bacteria (Table 4) with *Escherichia coli* ATCC 11303 proving to be the preferred organism. The aspartase activity of the *E. coli* displayed a lower pH optimum after entrapment (8.5 versus 10.5) and was insensitive to the addition of Mn^{2+} which normally stimulates this activity in free cells. These characteristics were elucidated using shake flasks containing 0.8 M ammonium fumarate solution which 0.8 mM Mn^{2+} added at 37° C for one hour. The observation that aspartase activity increased dramatically on prolonged use of the entrapped cells prompted the authors to test a variety of preparations as shown in Fig. 2. From the results obtained the researchers hypothesized that optimal enzyme activity was obtained after autolysis of the entrapped cells; a postulation corroborated by electron microscopy [7].

Table 4. Immobilization of various bacteria ([7], with permission from Applied Microbiology)

Bacterium	Aspartase activity of 1 ml of culture broth (μmol/hr)	Activity of immobilized cells per ml of culture broth (μmol/hr)	Yield of activity (%)
Bacterium succinium (IAM 1017)	15	4.3	29.0
Escherichia coli (ATCC 11303)	23	16.5	72.5
Proteus vulgaris (OUT 8226)	15	7.6	50.3
Pseudomonas aeruginosa (OUT 8252)	8	7.4	90.7
Serratia marcescens (OUT 8259)	12	5.2	41.0

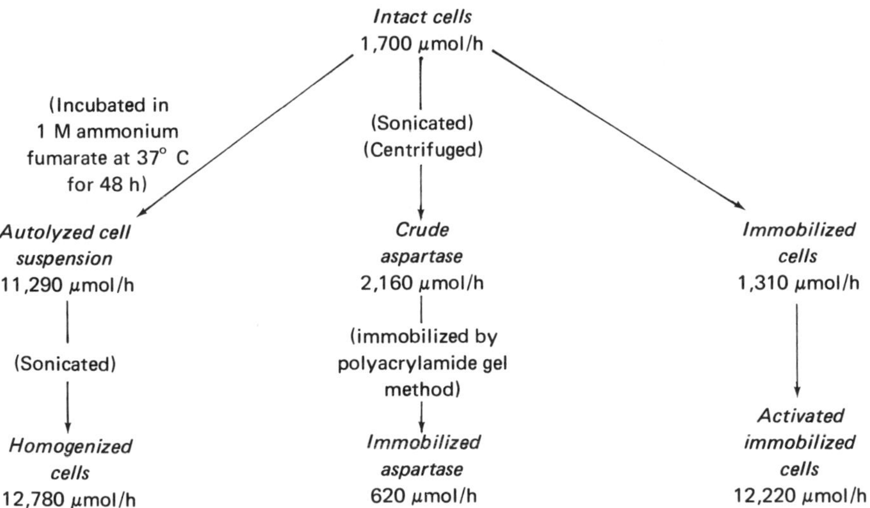

Fig. 2. Schematic comparison of aspartase activity of various enzyme preparations per unit of intact cells ([7] with permission from Applied Microbiology)

In a continuous column reactor, quantitative conversion of 1 M fumarate was achieved at 37° C pH 8.5 with a superficial velocity of 0.5 over an extended period of 36 days. Again, the loss of cell membrane integrity resulted in the leaching of intracellular metal ions and it proved necessary to supplement the feedstock with a divalent cation (1 mM Mn^{2+}, Mg^{2+}, or Ca^{2+}) to sustain enzyme activity. The L-aspartic acid was recovered from the effluent as a crystalline product in greater than 95% yields.

Under the same conditions, immobilized pure aspartase lost 75% of its activity after 20 days use. It is apparent that the natural environment provided by the whole cell immobilization technique does enhance the stability of this enzyme's activity.

A team of researchers from the laboratories of Tanabe Seiyaku Co., Ltd. [18] then developed a sectional packed column reactor to study the reaction mechanism and decay of aspartase activity for the immobilized whole cells of *E. coli* in polyacrylamide gel. The formation of L-aspartic acid proceeds as a zero order reaction, i.e. the reaction rate is independent of the concentration or diffusion rates of substrate or product in the range studied. The decay of the enzyme activity of the immobilized cells was exponential with time, 50% retention after four months operation at 37° C, occurred uniformly throughout the column and was independent of the volume of feedstock handled. An engineering analysis of the process was completed and an industrial reactor was designed and put into production in 1973. The overall production cost of this unit has been 60% of that for a conventional batch process using intact cells.

Considering the poor stability of the purified aspartase enzyme and the cost saving of a continuous column process over conventional batch processing, the important potential role of immobilized intact cells in industry becomes apparent.

Polyacrylamide encased cells of *Escherichia freundii* K-1 have an acid phosphatase activity capable of the transphosphorylation of glucose using p-nitrophenyl phosphate as a phosphate source. Continuous operation of the system in a column reactor gives glucose-6-phosphate as the major product along with a small amount of glucose-1-phosphate. The system is interesting in that the same enzyme catalyzes both the hydrolysis and phosphorylation steps. As a result, the flow rate in the column must be adjusted to maximize the hydrolysis of the p-nitrophenyl phosphate and to minimize the loss of the phosphorylated product. The column activity was retained with a 43% loss after 30 days and a 50% loss after 120 days of continuous operation [19].

Immobilized whole cells can be used to span the gap between single step conversions based on one enzyme activity and complex product formation involving considerable metabolic machinery.

Glutamic acid has been produced by gel entrapped cells of *Corynebacterium glutamicum* suspended in a glucose, mineral salts medium in a rotating flask at 30° C. The yield of product was studied as a function of repeated use, gel composition and storage. Although some loss of activity was noticed one re-use, significant activity was retained after storage of the gel entrapped cells for three months at 4° C. An unexpected observation was the enhancement of glutamic acid production by the polyacrylamide gel encased cells over that seen with free cells [20]. Attempts to adapt the system to a continuous column operation failed as a result of difficulties encountered in supplying the entrapped cells with oxygen [20].

The problem of supplying the entrapped cells with oxygen was also encountered by Martin and Perlman [9] in the oxidation of L-sorbose to L-sorbosone by *Gluconbacter melanogenus* IFO 3293. By molding "cell containing polyacrylamide films" of varying thicknesses, it was found that when the gel was over 0.7 mm thick L-sorbosone production was constant. Under the conditions of assay, cells lying deeper than 0.35 mm inside the polyacrylamide gel were being limited by oxygen deprivation. In a continuous column reactor, L-sorbosone production was proportional to the flow rate expected if dis-

solved oxygen were the limiting factor. Use of pure oxygen in aeration resulted in rapid loss of enzyme activity and attempts to use alternate electron acceptors did not enhance L-sorbosone production.
A further problem was the deactivation of the requisite enzyme system by the product, L-sorbosone. The only successful way found to stabilize the enzyme activity was the use of high concentrations of substrate in excess of the optimal conversion level. At 20% L-sorbose, the enzyme activity was stable for 15 days but, since the rate of production of L-sorbosone was constant above 3% L-sorbose, this led to poor conversion rates [9].
It is interesting to note that cell autolysis, toluene pretreatment of the cells or lyophilization of the cells before immobilization all enhanced the initial rate of L-sorbosone production by the immobilized system. It had been previously established that the same phenomenon occurred with free resting cells of *Gluconobacter melanogenus* and that the rate of production of L-sorbosone was a function of the rate of L-sorbose uptake. The pH and temperature maxima for this oxidation remained unchanged on immobilization of the cells in polyacrylamide gel [9].
Brevibacterium ammoniagenes IFO 12071 synthesizes coenzyme A in a multi-step process involving five enzymic steps; pantothenic acid → phosphopantothenic acid → phosphopantothenoylcysteine → phosphonatethiene → diphospho-CoA → CoA.
Polyacrylamide entrapment of the cells resulted in a somewhat greater heat stability, the same temperature optimum and a lower pH optimum (by one unit) than free dried cells with respect to their ability to accumulate coenzyme A. The gel entrapped cells were also stable to storage at 0–4° C for 45 days. Addition of sodium lauryl sulfate to the reaction mixture with gel entrapped cells enhanced coenzyme A accumulation but showed no such effect when added to gel entrapped cells which had been previously dried [8]. This behaviour is analogous to that observed above for the cetyltrimethylammonium bromide stimulation of urocanic acid production [15].
In a batch process, 7.1 g of dried cells entrapped in polyacrylamide gel gave 153 mg of isolated and purified coenzyme A when incubated 8 hrs at 37° C in a medium of sodium panthenate (0.5 μmol), cysteine (1 μmol), ATP (1.5 μmol), $MgSO_4$ (1 μmol) in 100 ml of phosphate buffer solution (pH 7.5) [8].
In a continuous column reactor, the production of coenzyme A fell to 50% of the original activity in 5 days. Considering the complexity of the conversion being effected, even this moderate longevity presents an encouraging demonstration of the potential utility of immobilized whole cells.

2.1.2 Collagen Membranes

Vieth *et al.* [21, 22] have developed an immobilization method which fixes whole cells in a reconstituted collagen membrane. Cells of *Streptomyces phaeochromogenes* and *Streptomyces venezuelae* immobilized in this way apparently retain glucose isomerase activity despite the severity of the treatment and operating conditions.
The cells which had been previously frozen were heated at 80° C for ninety minutes before use, presumably to prevent cell growth or division. The pretreated cells were then added to a collagen dispersion, the pH of which was changed from 6.5 to 11.5. This mixture was dried on a Mylar sheet to give a membrane (2–10 mm thick) which was

tanned in a 10% alkaline formaldehyde or glutaraldehyde solution for one half to five minutes and washed thoroughly with water. Chips of this material interspaced with a filter fabric were used as column packing for a continuous glucose to fructose isomerization process. The column operated continuously for forty days at 70° C displaying good stability until the spacer material began to break down. This occurred after fifteen days for undefined reasons. The access of the substrate and product to the cells was diffusion controlled [21, 22].

The harsh preparative treatment and high operating temperature of the column must result in great damage to the cells; yet the glucose isomerase activity remains. It would be reassuring to see evidence that the treated cells were indeed responsible for the observed activity.

2.1.3 Metal Hydroxide Precipitates

The addition of Ti^{4+} or Zr^{4+} chloride salts to water results in the pH dependent formation of gelatinous polymeric metal hydroxide precipitates in which the metals are bridged by hydroxyl or oxide groups. By conducting such a precipitation in a suspension of microbial cells (*Escherichia coli, Saccharomyces cerevisiae, Acetobacter sp.* or *Serratia marcescens*), the microorganisms have been entrapped in the gel-like precipitate formed. The viability of the contained cells was established in several ways. Entrapped *E. coli* or *Saccharomyces cerevisiae* cells were observed to take up oxygen at 30% the rate measured for equal numbers of free cells when incubated under appropriate conditions. Whether this reduced oxygen consumption resulted from cell damage, a decrease in the available cell surface area or from a diffusion limitation imposed by the gel was not shown. Further, when samples of gel-entrapped *Serratia marcescens* were used as an inoculum in fresh medium a large increase in the number of red cells was observed. Whether these cells were contained in the gel or released to the medium was not clearly stated and the identification of the new cells as *Serratia marcescens* was based on their visual appearance. The cells were reported as being firmly held by the gelatinous matrix with washings of the material being almost cell-free. Based on the known chemistry of metal hydroxide precipitates, Kennedy *et al.* [23] postulate the carboxyl, amino and hydroxyl groups on the surface of the cells undergo condensation reactions with unidentate metal hydroxide groups on the gel surface to give a covalent binding of the microbial cell to the gel. Since the gel formation is pH dependent, the choice of zirconium or titanium reagents is governed by the pH range required for the application of interest, titanium hydroxide being more effective in acidic conditions and zirconium hydroxide more effective at neutral or higher pH.

The continuous reactor adaptation of this system involved titanium (IV) hydroxide immobilized cells of *Acetobacter* which carried out the conversion of an alcoholic medium to acetic acid at a rate of 263 g/day (99% conversion). Unfortunately, further details as to the longevity, aeration problems, or operation of the process were omitted [23].

2.1.4 Agar Pellets

Whole cells of the yeast, *Saccharomyces pastorianus* were suspended in a 2.5% agar solution which was pelletized by direct injection of the hot solution (50° C) into cold toluene or tetrachloroethylene. The pellets formed were spherical with the cells distributed homogeneously throughout [24].

Sucrase activity was constant in a continuously fed fluidized bed at 47.5° C for over one hundred hours. Uptake of the invert sugar by the entrapped cells was negligible, i.e. < 0.1% of the sucrose hydrolyzed. Comparison of the stability of the cellular enzyme system with that of purified sucrase on DEAE-cellulose revealed that the whole cell system was superior.

The rate of hydrolysis by the agar entrapped yeast was correlated with the intraparticle enzyme concentration, the external substrate concentration and the pellet radius. This latter dependency arises from the diffusion barrier imposed by the agar. A consideration of mass transfer effects in this system and in another based on the lactase activity of encased *E. coli* have been published [24, 25].

2.1.5 Liquid Membrane Encapsulation

The sequential reduction of nitrate and/or nitrite by intact whole cells of *Micrococcus denitrificans* encapsulated in a liquid/surfactant membrane was demonstrated by Mohan and Li of the Exxon Research and Engineering Company [26, 27].

Encapsulation of cells was achieved by adding a phosphate buffered suspension of viable cells to a mixture of oil (86%), surfactant (2%), membrane strengthener (10%) and anion transport facilitator stirred at 600 rpm at 18° C. The emulsion so formed was then dispersed into a second phosphate buffered solution containing the nitrate and/or nitrite substrate. By this procedure, 500–600 cells end up entrapped in aqueous bubbles within oil droplets stabilized by the surfactant system as shown in Fig. 3.

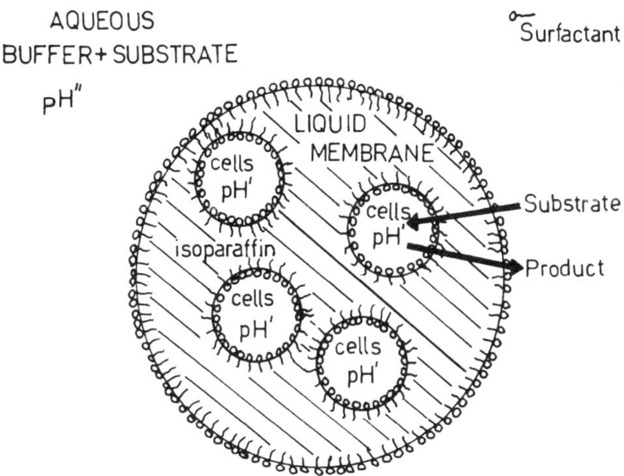

Fig. 3. Schematic representation of a liquid membrane microcapsule (20–40 μ diameter) containing *Micrococcus denitrificans* cells

Using batch flask conditions, the disappearance of the nitrate and nitrite substrate was monitored as a function of substrate concentration, cell concentration, pH of the internal buffer system (pH′ optimum, 7.6) and pH of the external phase (pH″ optimum, 6.0). The cells encapsulated in this way retained their enzymatic nitrate, nitrite reducing activity but the rate of reduction proved to be limited by the rate of substrate diffusion through the encapsulating system and consequently was dependent on the concentration of the anion transport facilitator (a high molecular weight secondary amine) added. Despite an inhibited rate of reduction relative to free cells, several advantages accrued from encapsulation. The entrapped cells were tolerant to a broad range of pH in the external phase (pH″) and retained activity in the external presence of 10^{-4} M $HgCl_2$ which totally inactivated the unbound cells. Further the longevity of the nitrate reducing activity was greatly enhanced by encapsulation. The system retained 78% of its activity after 120 hours compared to 0% for unbound cells after 16 hours. Mohan and Li have suggested that inclusion of nutrients with the cells may enhance the stability of the system even further [26, 27].

This study demonstrated that the encapsulation of *Micrococcus denitrificans* allows the use of a multi-enzyme reducing system which has an appreciable activity over extended periods (5.5 days at 24° C) and a tolerance to external pH variations and toxic substances. No co-factors or regenerating agents were required to maintain activity and the capacity for continuous operation (as required for application in secondary or tertiary waste water treatment) was demonstrated in a specially modified bio-reactor.

2.1.6 Miscellaneous

The amino acid acrylase activity of *Aspergillus oryzae* in cellulose nitrate has been the basis of a patent [28]. A patent has been issued for the covalent linking of cells by bifunctional agents [45] and for the flocculation of cells with polyelectrolytes [46].

A simple entrapment technique used to assay the enzyme activity of *Brevibacterium ammoniagenes* was the containment of the cells in cellophane tubing [8]. Entrapment in silastic resin [8], polyurea [7], or cross-linking with glutaraldehyde [8, 7] or 2,4-toluene diisocyanate [7] resulted in significant deactivation of the cellular enzymes under study.

2.1.7 Summary

The entrapment methods offer a straight-forward mechanical procedure for the immobilization of whole cells.

In liquid membrane encapsulation, collagen membrane inclusion or agar pelletization, a diffusion barrier is imposed on the operation of cellular enzyme systems. In some applications, this restriction on rate can be compensated by its beneficial side effects, such as the protection of the cells from variable or harsh external conditions.

The porosity of the polyacrylamide gel precludes these diffusion problems (unless oxygen is required) and this technique has consequently been the most popular for the industrial application of immobilized whole cell technology. There are certain microorganisms, however, in which enzymatic activity is lost in the polymerization procedure. In many instances, the permeability barrier imposed by the cell membrane is eliminated by autolysis, or lyophilization, drying or freezing of the cells or by treating the cells with toluene or surfactant. In these cases, cell viability is irrelevant to enzyme activity and may be lost.

No shedding of cells or progeny was reported for the continuous operation of systems using entrapped whole cells; this is a problem often encountered in adsorption or chemical immobilization methods as will be seen in the ensuing section.

2.2 Adsorption Techniques

2.2.1 Ion Exchange Materials

Ion exchange materials are established tools for the microbiologist (see the review by Rotman [29]) and have been employed in a variety of contexts including the removal of specific ions from microbial broth or media, and the thorough washing of cells reversibly bound to resin packed columns, the selective separation and concentration of specific species and the immobilization of microorganisms for continuous production of commercial materials, viral particles or synchronous cells.

From the data provided by the interaction of a large selection of bacteria and yeasts on both cationic and anionic exchange materials [29, 30] it is apparent that adsorption depends on the chemical nature of the cell wall surface with constituents such as peptides, diaminopimelic acid and hexosamine providing the necessary ionic sites for attachment to a charged support. The affinity of a specific species for a given ion exchanger is not predictable. There is no obvious reliable dependence on the bulk properties of the cell surface, such as the overall charge or electrophoretic mobility.

In many cases, this affinity is also dependent on the age of the cells (e.g., *E. coli* cells on aging develop an affinity for resins that previously held no attraction for the cells.) The explanation of cell binding by ion exchange materials must lie in a subtle balance of the number of surface charges, the specific configuration of charged sites and the accessibility of these ionic functional groups. In the above example, the suggestion is made that aging causes a change in the outer surface of the *E. coli* cells to make charged groups more accessible [29].

Of course, any factor which can affect surface charges can also influence this ionic based adsorption phenomenon. The presence of metal ions [30], or anions [31] can determine the degree of binding of cells to a cationic or anionic exchange resin by competition for the resin binding sites and by the neutralization or creation of cell surface charges. Indeed microorganisms are routinely eluted from such materials by the addition of acid or salt (see Table 5).

2.2.2 Chromatographic Separation of Microorganisms

The obvious application of such specific and manageable binding properties is the chromatographic separation of microbial mixtures or the concentration and isolation of a given microbe. Separations have been classified by Daniels and Kempe [31] into two types; (i) the preferential adsorption of one species of microbe over another which is not bound at all and, (ii) the indiscriminant adsorption of microbes followed by selective sequential elution of one species before another. Some representative examples of the technique are given in Table 5 [31, 32].

Table 5. The resolution of mixed suspensions of bacteria (based on material from Daniels and Kempe, 1967; Zvyagintsev and Gusev, 1971)

Species 1	Species 2	Exchange resin	Type of resolution
Escherichia coli	*Bacillus subtilis*	Anionic	Species 2 only adsorbed desorbed by low pH and addition of salt
Escherichia coli	*Staphylococcus aureus*	Anionic	Species 2 only adsorbed desorbed by low pH
Escherichia coli	*Pseudomonas ovalis*	Anionic	Species 2 only adsorbed desorbed by low pH
Staphylococcus aureus	*Bacillus subtilis*	Anionic	Both adsorbed species 1 desorbed by low pH species 2 desorbed by low pH and additions of salt
Bacillus subtilis	*Proteus vulgaris*	Cationic	Species 2 only adsorbed desorbed by high pH
Bacillus subtilis	*Bacillus cereus*	Anionic	Both adsorbed Species 1 desorbed by low salt Species 2 desorbed by high salt

2.2.3 Properties and Applications of Adsorbed Microorganisms

It has been shown that the growth of bacteria on the walls of a continuous-culture vessel can significantly influence the population density and growth kinetics of the suspended cells. In studies on predator-prey relationships where the grazing of protozoa on bacteria depends on engulfment of suspended bacteria, the presence of a "reservoir" of bacterial cells bound to the chemostat walls significantly alters the population dynamics of the system [33, 34].

On examination of several genera, Larsen and Dimmick [5] noted that for the species, *Serratia marcescens* and *Escherichia coli,* up to 90% of the cells in suspension had been spawned by an almost invisible wall bound population. In other cases, notably from the genus *Bacillus,* no such contribution was found. Helmstetter and Cooper [35, 36] reported that *E. coli B,* adsorbed on a filter membrane in flowing nutrient medium, had a shorter doubling time than the freely suspended microorganism. Obviously, any change in the metabolism of the adsorbed cell population will have profound influence on continuous-culture processes.

Physiological alteration of cells on adsorption to Dowex I anionite occurs for *E. coli* and *Azotobacter agile* ATCC 9046. The effect appeared to be three-fold, involving: the increase of the pH optimum for growth or substrate oxidation by one unit on adsorption, the reduction of the rate of oxidation of glucose or succinate and a greatly reduced lag time in the induced oxidation of succinate or citrate by glucose grown cells, a trait retained by the cells on desorption [37, 38]. Changes in the pH optimum were attributed to the influence of the cations present at the surface of the anion exchange resin. By preventing cell division through the use of chloramphenicol or the omission of a carbon or nitrogen source from the medium, it was possible to discriminate between *E. coli Yamaguchi* cells released by desorption and those released as progeny from cell division. This secondary growth was studied by continuous growth of the suspended immobilized cells at a dilution rate sufficient to "wash out" free cells. The free cell concentration in the effluent of the experiment displayed, at least initially, a periodicity and it was found that effluent cells would divide synchronously in growth media at pH 7–8. At least initially then, release of progeny by cell division from the resin surface proceeded according to the age distribution of the adsorbed cells.

In both batch studies and continuous cultivation, the adsorbed cells had a shortened generation time which paradoxically is correlated with a lowered chemical activity and a reduced RNA content as measured on the freshly desorbed cells [39–41].

The release of progeny from a population of adsorbed viable whole cells prevents the use of this system for a continuous production process since where enzymatic activity correlates with viability the effluent from a column based process would be constantly seeded with released cells.

Johnson and Ciegler [42] therefore have used a column packed with a cellulose functionalized with mixed amine groups, ECTEOLA-cellulose, on which fungal spores of *Aspergillus wentii* NRRL 2001, *Aspergillus oryzae* NRRL 1989 and *Penicillium roqueforti* NRRL 3360 were adsorbed. Spores, while physiologically inert, do display selective enzymatic activity. An attempt was made to follow starch hydrolysis by *A. wentii* spores, fatty acid oxidation by *P. roqueforti* spores and invertase activity by *A. oryzae* spores [42].

Unfortunately, the first of these processes proved too slow and the second too difficult to monitor. The hydrolysis of sucrose by *A. oryzae* spores however proved amenable to study in a continuous column system. A direct dependence of the extent of hydrolysis on spore concentration was observed as expected. One interesting feature of the work was the use of gangs of short columns to improve the overall conversion from 15–20% for a single solumn to 30–35% for two columns in sequence.

In all instances, the columns proved to be stable and germination of the spores was negligible provided the columns were washed free of sucrose before storage or after 8–10 hours continual use. Alternately, the inclusion of 1 mM iodoacetic acid in the feedstock sucrose solution prevented germination without affecting the invertase activity.

In the Johnson and Ciegler study [42], a number of potential support materials was surveyed before selecting ECTEOLA-cellulose. Silica gel, Amberlite IR-120 and neutral cellulose were unsuitable due to poor spore retention and flow characteristics. An ion exchange polyacrylamide gel and Carboxymethyl-Biogel also failed to retain spores adequately. Of the celluloses tried (carboxymethyl-phosphonic acid, diethylaminoethyl- and ECTEOLA- [mixed amines]), the ECTEOLA-cellulose was preferred on the basis of its spore retention and flow rate properties.

On the whole, the lability of the microorganisms adsorbed to ion exchange materials precludes their use in continuous industrial processes.

2.3 Selective Binding of Cells by Immobilized Macromolecules

In 1969, Cautrecasas found that immobilized insulin would oxidize glucose in the presence of isolated fat cells at rates comparable to free soluble insulin. The implication is that the "principal interaction of insulin is with the cell surface structures and that the structural requirements for this interaction are essentially intact in the insulin-agarose derivative" [43, 44].

The idea that macromolecules immobilized on a support could retain their ability to bind certain specific sites on the surface of suspended cells led to a new method of whole cell immobilization.

The macromolecules used are lectins, naturally occurring proteins that have the ability to agglutinate certain kinds of cells. They are commonly extracted from plants, especially legumes, and from other sources, such as snails or fish. Their agglutinating activity arises from their ability to bind specific antigens on the surface of cells, thus producing a clumping effect. By attaching these macromolecules to an inert support system, it is therefore, possible to selectively bind whole cells. For example, wheat germ agglutinin will preferentially agglutinate malignant animal cells in a mixed cell suspension with normal cells. The process is inhibited or reversed by N-acetylglucosamine which is thought to compete for the binding site of the wheat germ agglutinin on the cell surface.

By covalently binding the wheat germ agglutinin to polyacrylamide beads using standard protein immobilization techniques, it was found that mouse leukemia (L 1210) cells could be selectively adsorbed. Care was required in binding the protein to the polyacrylamide since the presence of too many covalent links distorted the tertiary structure of the agglutinin and destroyed its binding capacity [47].

Comparative studies of the binding of lymphoma, myeloid leukemia and normal fibroblast cells to Concanavalin A or wheat germ agglutinin immobilized on nylon fibres has used as a basis for the study of cell binding induced by plant lectins and Concanavalin A [48].

Lectin from *Lens culinaris* covalently linked to 2B-Sepharose beads will readily bind HeLa cells or SV3T3 cells in the absence of hapten sugars. In the presence of 0.2 M methyl-α-D-glucopyranoside or methyl-α-D-mannopyranoside however, this binding is diminished (or even eliminated in the latter case) by the competition of these sugars with the cells for the lectin binding sites [49]. The binding of cells to immobilized lectin can be reversed within one hour by the addition of a hapten sugar solution. The bound cells were shown to exclude trypan blue dye, a sign of viability, and could multiply after their sugar promoted release from the lectin coated support [49]. The use of lectin, however, allows the binding and release of cells under physiological conditions sufficient to maintain cell viability [49].

In a similar approach, the selective immunoadsorption of cells on an antibody-coated inert support allows the chromatographic removal or isolation of cells having specific immune activity. Examples of the technique include the binding of mouse cells to glass or plastic beads [50], the binding guinea pig erythrocytes to an open cell polyurethane foam [51] and mouse spleen cells to polyacrylamide beads [52].

The use of immobilized macromolecules to bind whole cells has to date been limited to the adsorption of cells of animal origin. Nevertheless, one important instance of lectin promoted binding of whole bacterial cells to a surface has been found in nature. The fixation of nitrogen by legumes depends on the invasion of the plant root hairs by a symbiotic bacteria. This process is highly species specific and evidence has been found by Dazzo and Hubbel [53] that the initial binding of infective strains of *Rhizobium trifolii* to the surface of clover root hairs on *Trifolium repens* and *Trifolium fragiferum* takes place via a species specific clover lectin complex which bridges the cross-reactive antigens on the surface of the root hairs and on the surface of the bacterial cell. This natural example demonstrates the existence of lectins with selective affinity for bacterial cells and, presumably, the immobilization of these lectins would result in a highly selective binding of bacterial cells.

These relatively sophisticated modes of whole cell immobilization are as yet without practical application.

2.4 Covalent or Coordinate Bonding of Cell to Support

The immobilization of whole cells to a support via irreversible covalent bonds could produce a system free of the diffusion limitations present in an entrapment procedure and free of the release of progeny typical of the adsorption systems.

Cell wall fragments of *Bacillus subtilis Marburg* and whole cells of *E. coli B/R, Staphylococcus aureus, Pseudomonas aeruginosa* and *B. subtilis Marburg* have been immobilized by Murray and Beveridge, with varying degrees of success, using the carbodiimide assisted formation of covalent linkages between the cells and agarose adipic hydrazide beads [54]. Unfortunately, carbodiimide reagents are highly toxic and activation of the carboxylate groups on the cell surface with these reagents results in the loss of both the enzyme activity and viability of the cell [55].

By reversing the procedure and activating the carboxylate groups on the appropriate agarose beads such as Affigel 201 with the carbodiimide, it is possible to immobilize *B. subtilis Marburg* or *Micrococcus luteus* cells in a two step process which avoids exposure of the cells to carbodiimide [56].

The *Micrococcus luteus* system has also been immobilized on carboxymethylcellulose and although inviable retains its histidine ammonia-lyase activity. The production of urocanic acid from histidine in a continuous column reactor is currently under study [56].

The promising results obtained with this system suggest that reactive functional groups are available on the cell surface and that the standard methods of covalent or coordinate bonding of materials to specially prepared supports are feasible as modes of whole cell binding.

One attempt to covalently bind *Brevibacterium ammoniagenes* IFO 12071 to a copolymer of ethylene-maleic anhydride has been reported and resulted in the loss of the coenzyme A synthesizing activity of the cells [8].

3. Conclusion

The literature of the last three years has seen a renaissance in the use of immobilized enzyme systems made practical by the immobilization of the entire cell. Such systems have proven to have a diversity of application unattainable with isolated pure enzymes and the operation of such systems in continuous reactors has proven to be less costly than more conventional processes based on free intact cells. Thus, whole cell immobilization as a technique should be of expanding practical and academic significance in the future.

References

1. Zaborsky, O.: Immobilized Enzymes, CRC Press, Cleveland (1973).
2. Newman, H. N.: Microbios **9**, 247 (1974).
3. Zvyagintsev, D., Pertososkaya, A. F., Yakhnin, E. D., Averbach, E. I.: Mikrobiologiya **40**, 1024 (1971).
4. Atkinson, B., Knight, A. J.: Biotechnol. Bioeng. **17**, 1245 (1975).
5. Larsen, D. H., Dimmick, R. L.: J. Bac. **88**, 1381 (1964).
6. Van Wezel, A. L.: In: Tissue Culture Methods and Applications. (Patterson, M. K., Kruuse Jr., P., eds.), p. 372. New York: Academic Press 1973.
7. Chibata, I., Tosa, T., Sato, T.: Applied Microbiol. **27**, 878 (1974).
8. Shimizu, S., Morioka, H., Tani, Y., Ogita, K.: J. Ferment. Technol. **53**, 77 (1975).
9. Martin, G. K. A., Perlman, D.: Biotechnol. Bioeng. **18**, 217 (1976).
10. Mosbach, K., Mosbach, R.: Acta Chem. Scand. **20**, 2807 (1966).
11. Updike, S. M., Harris, D. R., Shrago, E.: Nature (London) **224**, 1122 (1969).
12. Mosbach, K., Larsson, P. O.: Biotechnol. Bioeng. **12**, 19 (1970).
13. Franks, N. E.: Biochem. Biophys. Acta **252**, 246 (1971).
14. Franks, N. E.: In: Enzyme Engineering (Wingard jr., L. B., ed.), p. 327. New York: Wiley Interscience 1972. (Biotechnol. Bioeng., Symp. No. 3).
15. Yamamoto, K., Sato, T., Tosa, T., Chibata, I.: Biotechnol. Bioeng. **16**, 1601 (1974).
16. Yamamoto, K., Sato, T., Tosa, T., Chibata, I.: Biotechnol. Bioeng. **16**, 1589 (1974).
17. Tosa, T., Sato, T., Mori, T., Chibata, I.: Applied Microbiol. **27**, 886 (1974).
18. Sato, T., Mori, T., Tosa, T., Chibata, I., Furui, M., Yamashita, K., Sumi, A.: Biotechnol. Bioeng. **17**, 1797 (1975).
19. Saif, S., Tani, Y., Ogata, K.: J. Ferment. Technol. **53**, 380 (1975).
20. Slowinski, W., Cham, S. E.: Biotechnol. Bioeng. **15**, 973 (1973).
21. Vieth, W. R., Wang, S. S., Saini, R.: Biotechnol Bioeng. **15**, 565 (1973).
22. Venkatasubramanian, K., Saini, R., Vieth, W. R.: J. Ferment. Technol. **52**, 268 (1974).
23. Kennedy, J. F., Barker, S. A., Humphreys, J. D.: Nature (London) **261**, 242 (1976).
24. Toda, K., Shoda, M.: Biotechnol. Bioeng. **17**, 481 (1975).
25. Toda, K.: Biotechnol. Bioeng. **17**, 1729 (1975).
26. Mohan, R. R., Li, N. N.: Biotechnol. Bioeng. **17**, 1137 (1975).
27. Mohan, R. R., Li, N. N.: Biotechnol. Bioeng. **16**, 513 (1974).
28. Leuschner, F.: German Pat. 1227855 (Cl. C12d).
29. Rotman, B.: Bac. Rev's. **24**, 251 (1960).
30. Zvyagintsev, D. G.: Mikrobiologiya **31**, 339 (1971).
31. Daniels, S. L., Kempe, L. L.: In: Chemical Engineering in Medicine and Biology (Herschey, D., ed.), p. 391. New York: Plenum Press 1967.

32. Zvyagintsev, D. G., Gusev, V. S.: Mikrobiologiya **40**, 139 (1971).
33. Bonomi, A., Fredrickson, A. G.: Biotechnol. Bioeng. **18**, 239 (1976).
34. Van den Ende, P.: Science **181**, 562 (1973).
35. Helmstetter, C. E.: J. Mol. Biol. **24**, 417 (1967).
36. Helmstetter, C. E., Cooper, S.: J. Mol. Biol. **31**, 507 (1968).
37. Hattori, T., Furusaka, C.: Biochem. (Tokyo) **50**, 312 (1961).
38. Hattori, T., Furusaka, C.: Biochem. (Tokyo) **48**, 831 (1960).
39. Hattori, R., Hattori, T., Furusaka, C.: J. Gen. Appl. Microbiol. **18**, 271 (1972).
40. Hattori, R., Hattori, T., Furusaka, C.: J. Gen. Appl. Microbiol. **18**, 285 (1972).
41. Hattori, R.: J. Gen. Appl. Microbiol. **18**, 319 (1972).
42. Johnson, D. E., Ciegler, A.: Arch. Biochem. and Biophys. **130**, 384 (1969).
43. Cautrecasas, P.: Proc. Nat. Acad. Sci., U.S. **63**, 450 (1969).
44. Cautrecasas, P.: In: Biochemical Aspects of Reactions on Solid Support (Stark, G. R., ed.), p. 79. New York: Plenum Press 1971.
45. Moskowitz, G. J.: U.S. Patent 3843442 (1974).
46. Lee, C. K., Long, M. E.: U.S. Patent 3821086 (1974).
47. Zabriskie, D., Ollis, D. F., Burger, M. M.: Biotechnol. Bioeng. **15**, 981 (1973).
48. Rutishauser, U., Sachs, L.: J. Cell Biol. **65**, 247 (1975).
49. Kinzel, V., Kubler, D., Richards, J., Stohr, M.: Science **192**, 487 (1976).
50. Wigzell, H., Anderson, B.: J. Exp. Med. **129**, 23 (1969).
51. Evans, W. H., Mage, M. G., Peterson, E. A.: J. Immunol. **102**, 899 (1969).
52. Truffi-Bachi, P., Wofsy, L.: Proc. Nat. Acad. Sci., U.S. **66**, 685 (1970).
53. Dazzo, F. D., Hubell, P. H.: Applied Microbiol. **30**, 1017 (1975).
54. Beveridge, T., Murray, R. G. E.: Private Communication.
55. Chipley, J. R.: Microbios **10**, 115 (1974).
56. Jack, T. R., Zajic, J. E.: Biotechnol. Bioeng., in press.

Related Titles

Vol. 19 **Polymerization Reactions**
37 figures. 24 tables, III, 164 pages. 1975.
ISBN 3-540-07460-0
A. Ledwith; D. C. Sherrington, Stable Organic Cation Salts: Ion Pair Equilibria and Use in Cationic Polymerization (167 references).
J. P. Kennedy; J. E. Johnston, The Cationic Isomerization Polymerization of 3-Methyl-1-butene and 4-Methyl-1-pentene (20 references).
E. B. Mano; F. M. B. Coutinho, Grafting on Polyamides (194 references).
F. Millich, Rigid Rods and the Characterization of Polyisocyanides (37 references).

Vol. 20 **New Scientific Aspects**
95 figures. IV, 227 pages. 1976.
ISBN 3-540-07361-X
Y. Imanishi, Syntheses, Conformation and Reactions of Cyclic Peptides (166 references).
E. Iizuka, Properties of Liquid Crystals of Polypeptides with Stress on the Electromagnetic Orientation (93 references).
J. Sohma; M. Sakaguchi; ESR Studies on Polymer Radicals produced by Mechanical Destruction and their Reactivity (104 references).
T. Kunitake; Y. Okahata, Catalytic Hydrolysis by Synthetic Polymers (146 references).

Vol. 21 **Mechanisms of Polyreactions – Polymer Characterization**
68 figures. III, 151 pages. 1976.
ISBN 3-540-07727-8
J. P. Kennedy; T. Chou; Poly (isobutylene-co-β-Pinene): A New Sulfur Vulcanizable, Ozone Resistant Elastomer by Cationic Isomerization Copolymerization (55 references).
J. A. Semlyen, Ring-Chain Equilibria and the Conformations of Polymer Chains (224 references).
S. Inoue, Asymmetric Reactions of Synthetic Polypeptides (67 references).
J.-M. Braun, J. E. Guillet, Study of Polymers by Inverse Gas Chromatography (108 references).

Vol. 22 **Physical Chemistry**
77 figures. IV, 154 pages
ISBN 3-540-07942-4
Y. S. Lipatov, Relaxation and Viscoelastic Properties of Heterogeneous Polymeric Compositions (171 references).
B. R. Jennings, Electro-Optic Methods for Characterizing Macromolecules in Dilute Solution (32 references).
A. M. Basedow; K. Ebert, Ultrasonic Degredation of Polymers in Solution (100 references).
Scheduled to come out in 1977.

Springer-Verlag Berlin · Heidelberg · New York